Algrove Publishing Limited
36 Mill Street, P.O. Box 1238
Almonte, Ontario, Canada K0A 1A0

Telephone: (613) 256-0350
Fax: (613) 256-0360
Email: sales@algrove.com

Credit: *The 1905 volume used for reproduction came from the library of the late Jack Bruce of Calgary. We are grateful to his widow, Dorothy Bruce, for making it available to us.*

Library and Archives Canada Cataloguing in Publication

Musson's improved lumber and log pocket book.

(Classic reprint series)
Reprint. First published as: Musson's improved lumber and log book,
1905. Toronto : Musson, [1905?].
ISBN 1-897030-48-7

1. Lumber trade--Tables. 2. Ready-reckoners. I. Title: Improved lumber and log pocket book. II. Series: Classic reprint series (Almonte, Ont.)

TS847.M87 2006   674.02'12   C2006-901710-7

Printed in Canada
#1-4-06

PUBLISHER'S NOTE

A century ago loggers had far less education than today and nobody had electronic calculators. Books like this were a necessity in the lumber business.

This particular book contains much more advice than was usual for these guides. A few bits of the advice are suspect (e.g., the medicinal qualities of turpentine) but there is much technical material that is of interest and would not be found in any literature on the subject today.

Leonard G. Lee, Publisher
Almonte, Ontario
April 2006

WARNING

This is a reprint of a publication compiled in 1905. It describes what was done and what was recommended to be done in accordance with the knowledge of the day. It should be read in this light.

It is advisable to treat all corrosive, explosive or toxic materials with greater caution than is indicated here, particularly any materials that come in contact with the body. Treat all chemicals with respect and use them only in ways sanctioned by the sellers and the law.

Musson's *IMPROVED* LUMBER AND LOG POCKET BOOK

1905

FOR SHIP AND BOAT BUILDERS, LUMBER
MERCHANTS, SAW-MILL MEN, FARMERS
AND MECHANICS

❧

BASED ON

J. M. SCRIBNER'S LOG BOOK

Author of "Engineers' and Mechanics' Companion," "Engineers' Table Book," Etc.

AND ON DOYLE'S RULE

❧

REVISED ILLUSTRATED EDITION

❧

TORONTO

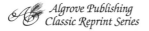

Algrove Publishing
Classic Reprint Series

THIS EDITION

Contains all the tables used in former editions. Also thirty pages of Musson's addition of new Wholesale Tables

SAW-LOGS REDUCED
— TO —
INCH BOARD MEASURE
BY DOYLE'S RULE

Stave and Heading Bolt Tables, Cord Wood, Prices of Lumber per foot, Speed of Circular Saws, Weights of Wood, Strength of Ropes, Felling of Trees, Growth of Trees, Cost of Fences, Price of Standard Logs, &c., &c.

Among the vast number of recommendations which we have received from time to time, we think it unnecessary to insert any here, as the book is too well known to require their publication. The large sale of this book is a sufficient evidence of its popularity.

CONTENTS

CONTENTS

6 CONTENTS

PREFACE

Scribner's Lumber and Log Tables having been published for nearly fifty years, we now present the fifth revision of this work.

The best features of Doyle's rules have been incorporated with Scribner.

The present edition of 1905, contains *forty pages* of new tables, miscellaneous matter and illustrations not included in former editions.

SCRIBNER'S LUMBER AND LOG BOOK long since won for itself more than a national reputation.

Over a *million and a half copies* have been sold in the United States and Canada, while extensive orders have been received from Europe, Central and South America, Mexico and Australia.

We submit the present edition of this justly popular book to the public, confident that it will continue to be recognized as the STANDARD LUMBER AND LOG BOOK.

THE PUBLISHER.

Toronto, 1905.

MULTIPLICATION TABLE

THIS IS INSERTED FOR THOSE WHO HAVE NOT THOROUGH-LY COMMITTED IT TO MEMORY

1	2	3	4	5	6	7	8	9	10	11	12
2	4	6	8	10	12	14	16	18	20	22	24
3	6	9	12	15	18	21	24	27	30	33	36
4	8	12	16	20	24	28	32	36	40	44	48
5	10	15	20	25	30	35	40	45	50	55	60
6	12	18	24	30	36	42	48	54	60	66	72
7	14	21	28	35	42	49	56	63	70	77	84
8	16	24	32	40	48	56	64	72	80	88	96
9	18	27	36	45	54	63	72	81	90	99	108
10	20	30	40	50	60	70	80	90	100	110	120
11	22	33	44	55	66	77	88	99	110	121	132
12	24	36	48	60	72	84	96	108	120	132	144

	PER HOUR	PER SEC.
A man travels..................	3 miles....	4 feet
A horse trots..................	7 "	.... 10 "
A horse runs..................	20 "	.... 29 "
Steamboats run..................	18 "	.... 26 "
Sailing vessels run..............	10 "	.... 14 "
Slow rivers flow................	3 "	.... 4 "
Rapid rivers flow................	7 "	.... 10 "
A moderate wind blows........	7 "	.... 10 "
A storm moves................	36 "	.... 52 "
A hurricane moves..............	80 "	.... 117 "
A rifle ball moves.............. 1000	"	....1466 "

HINTS TO LUMBER DEALERS AND MECHANICS IN SELECTING MATERIALS FOR BUILDING PURPOSES

SELECTION OF STANDING TREES

The principal circumstances which affect the quality of growing trees, are *soil, climate* and *aspect*.

In a moist soil, the wood is less firm, and decays sooner than in a dry, sandy soil; but in the latter, the timber is seldom fine; the best is that which grows in a dark soil, mixed with stones and gravel. This remark does not apply to the poplar, willow, cypress and other light woods, which grow best in wet situations.

In the United States, the climate of the Northern and Middle States is most favorable to the growth of timber used for ordinary purposes, except the cypress.

Trees growing in the centre of a forest, or on a plain, are generally straighter and more free from limbs than those growing on the edge of the forest, in open ground, or on the sides of hills; but the former are at the same time less hard; the toughest part of a tree will always be found on the side next the north.

The aspect most sheltered from the prevalent winds is generally most favorable to the growth of timber. The vicinity of salt water is favorable to the strength and hardness of white oak.

The selection of timber trees should be made before the fall of the leaf. A healthy tree is indicated by the top of branches being vigorous, and well covered with leaves; the bark is clear, smooth, and of a uniform color. If the top has a regular, rounded form—if the bark is dull, scabby, and covered with white and red spots, caused by running water or sap—the tree is unsound. The decay of the uppermost branches, and the separation of the bark from the wood, are infallible signs of the decline of a tree.

DEFECTS OF TIMBER TREES (ESPECIALLY OF OAK)

SAP, the white wood next to the bark, which very soon rots, should never be used, except that of hickory. There are sometimes found rings of light-colored wood surrounded by good hard wood, this may be called the *second sap;* it should cause the rejection of the tree.

BRASH-WOOD is a defect generally consequent on the decline of the tree from age; the pores of the wood are open, the wood is reddish colored, it breaks short, without splinters, and the chips crumble to pieces. This wood is entirely unfit for mechanical purposes or artillery carriages.

WOOD WHICH HAS DIED BEFORE BEING FELLED should in general be rejected; so should *knotty trees*, and those which are covered with tubercles, &c.

TWISTED WOOD, the grain of which ascends in a spiral form, is unfit for use in large scantling; but if the defect is not very decided, the wood may be used for naves, and for some light pieces.

SPLITS, CHECKS AND CRACKS, extending towards the centre, if deep and strongly marked, make the wood unfit for use, unless it is intended to be split.

WIND-SHAKES are cracks separating the concentric layers of wood from each other; if the shake extends through the entire circle, it is a ruinous defect.

All the above mentioned defects are to be guarded against in procuring timber for use in artillery constructions; the *center heart* is also to be rejected in nearly all cases.

FELLING TIMBER

The most suitable season for felling timber, is that in which vegetation is at rest, which is the case in mid-winter and in mid-summer; recent opinions, derived from facts, incline to give preference to the latter season, say the month of July; but the usual practice is to fell trees for timber between the first of December and middle of March. Some experiments are in progress with a view to determine the question with regard to oak timber for ordinary purposes.

The tree should be allowed to obtain its full maturity before being felled; this period in oak timber is generally at the age of from 75 to 100 years, or upwards, according to circumstances. The age of hard wood is determined by the number of rings which may be counted in a section of the tree.

The tree should be cut as near the ground as possible, the lower part being the best timber. The quality of the wood is in some degree indicated by the color, which should be nearly uniform in the heart wood, a little deeper toward the center, and without sudden transitions.

Felled timber should be immediately stripped of its bark, and raised from the ground.

As soon as practicable after the tree is felled, the sap-wood should be taken off, and the timber reduced, either by sawing or splitting, nearly to the dimensions required for use.

The best method of preventing decay is the immediate removal of it to a dry situation, where it should be piled in such a manner as to secure a free circulation of air around it, but without exposure to the sun and wind. When thoroughly seasoned, before cutting it up into smaller pieces, it is less liable to warp and twist in drying.

When green, timber is not so *strong* as when thoroughly dry.

Lumber containing much sap is not only weaker but decays much sooner than that free from sap.

SEASONING AND PRESERVING TIMBER

For the purpose of seasoning, timber should be piled under shelter, where it may be kept dry, but not exposed to a strong current of air; at the same time, there should be a free circulation of air about the timber, with which view slats or blocks of wood should be placed between the pieces that lie over each other, near enough to prevent the timber from bending.

In the sheds, the pieces of timber should be piled in this way, or in square piles, and classed according to age and kind. Each pile should be distinctly marked with the number and kind of pieces, and the age, or the date of receiving them.

The piles should be taken down and made over again at intervals, varying with the length of time which the timber has been cut.

The seasoning of timber requires from two to four years, according to its size.

Gradual drying and seasoning in this manner is considered the most favorable to the durability and strength of timber, but various methods have been prepared for hastening the process. For this purpose, *steaming* and *boiling* timber has been applied with success; *kiln-drying* is serviceable only for boards and pieces of small dimensions, and is apt to cause cracks, and to impair the strength of wood, unless performed very slowly.

Timber of large dimensions is improved by *immersion in water* for some weeks, according to its size, after which, it is less subject to warp and crack in steaming.

Oak timber loses about *one-fifth of its weight* in seasoning, and about *one-third of its weight* in becoming dry.

DURABILITY OF DIFFERENT WOODS

Experiments have been lately made by driving sticks, made of different woods, each two feet long and one and one-half inches square, into the ground, only one-half an inch projecting outward. It was found that in five years, all those made of oak, elm, ash, fir, soft mahogany, and nearly every variety of pine, were totally rotten. Larch, hard pine and teak wood were decayed on the outside only; while acacia, with the exception of being also slightly attacked on the exterior, was otherwise sound. Hard mahogany and cedar of Lebanon were in tolerably good condition; but only Virginia cedar was found as good as when put in the ground. This is of some importance to builders, showing what woods should be avoided, and what others used by preference in underground work.

The duration of wood when kept dry, is very great, as beams still exist which are known to be nearly 1,100 years old. Piles driven by the Romans prior to the Christian era, have been examined of late, and found to be perfectly sound after an immersion of nearly 2,000 years.

The wood of some tools will last longer than the metals, as in spades, hoes and ploughs. In other tools the wood is first gone, as in wagons, wheelbarrows and machines. Such wood should be painted or oiled; the paint not only looks well but preserves the wood; Petroleum oil is as good as any other.

Hard wood stumps decay in five to six years; spruce stumps decay in about the same time; hemlock stumps in eight to nine years; cedar eight to nine years; pine stumps, never.

Cedar, oak, yellow pine and chestnut are the most durable woods in dry places.

LOADING LOGS ON A WAGON—THE CUT EXPLAINS ITSELF.

SCANTLING MEASURE

Accurately Reduced to Board Measure

EXPLANATION

The length of any piece of scantling or timber will be found in the left hand column, under the side dimensions. The breadth and depth (or side dimensions), in inches, will be found at the head of each column of computations. Thus, on page 19, a piece of scantling 2½ by 11 inches, side dimensions, and 16 feet long, is shown to contain 36 feet and 8 inches of board measure. On page 21 a piece of scantling 4 by 10 inches, side dimensions, and 17 feet long, is shown to contain 56 feet 8 inches, board measure. The answer sought for in all cases, will be found directly on the right of the length, and under the side dimensions. If a piece of scantling, or stick of timber, should exceed, in length, any provision which has been made in these tables, its contents would be shown by taking twice what is given for half its length. Thus, a piece of scantling 46 feet long, would contain twice as many feet, board measure, as is shown in the table to be the contents of a stick 23 feet long. So, also, one 39 feet long would contain as many feet, board measure, as these tables show opposite to 22 and 17 feet long, or three times the contents of one 13 feet long.

SCANTLING MEASURE

2 x 2		2 x 3		2 x 4		2 x 5		2 x 6	
Length 1	0.4	Length 1	0.6	Length 1	0.8	Length 1	0.10	Length 1	1.
2	0.8	2	1.	2	1.4	2	1.8	2	2.
3	1.	3	1.6	3	2.	3	2.6	3	3.
4	1.4	4	2.	4	2.8	4	3.4	4	4.
5	1.8	5	2.6	5	3.4	5	4.2	5	5.
6	2.	6	3.	6	4.	6	5.	6	6.
7	2.4	7	3.6	7	4.8	7	5.10	7	7.
8	2.8	8	4.	8	5.4	8	6.8	8	8.
9	3.	9	4.6	9	6.	9	7.6	9	9.
10	3.4	10	5.	10	6.8	10	8.4	10	10.
11	3.8	11	5.6	11	7.4	11	9.2	11	11.
12	4.	12	6.	12	8.	12	10.	12	12.
13	4.4	13	6.6	13	8.8	13	10.10	13	13.
14	4.8	14	7.	14	9.4	14	11.8	14	14.
15	5.	15	7.6	15	10.	15	12.6	15	15.
16	5.4	16	8.	16	10.8	16	13.4	16	16.
17	5.8	17	8.6	17	11.4	17	14.2	17	17.
18	6.	18	9.	18	12.	18	15.	18	18.
19	6.4	19	9.6	19	12.8	19	15.10	19	19.
20	6.8	20	10.	20	13.4	20	16.8	20	20.
21	7.	21	10.6	21	14.	21	17.6	21	21.
22	7.4	22	11.	22	14.8	22	18.4	22	22.
23	7.8	23	11.6	23	15.4	23	19.2	23	23.
24	8.	24	12.	24	16.	24	20.	24	24.
25	8.4	25	12.6	25	16.8	25	20.10	25	25.
26	8.8	26	13.	26	17.4	26	21.8	26	26.
27	9.	27	13.6	27	18.	27	22.6	27	27.
28	9.4	28	14.	28	18.8	28	23.4	28	28.
29	9.8	29	14.6	29	19.4	29	24.2	29	29.
30	10.	30	15.	30	20.	30	25.	30	30.

SCANTLING MEASURE

2 x 7		2 x 8		2 x 9		2 x 10		2 x 11	
Length 1	1.2	Length 1	1.4	Length 1	1.6	Length 1	1.8	Length 1	1.10
2	2.4	2	2.8	2	3.	2	3.4	2	3.8
3	3.6	3	4.	3	4.6	3	5.	3	5.6
4	4.8	4	5.4	4	6.	4	6.8	4	7.4
5	5.10	5	6.8	5	7.6	5	8.4	5	9.2
6	7.	6	8.	6	9.	6	10.	6	11.
7	8.2	7	9.4	7	10.6	7	11.8	7	12.10
8	9.4	8	10.8	8	12.	8	13.4	8	14.8
9	10.6	9	12.	9	13.6	9	15.	9	16.6
10	11.8	10	13.4	10	15.	10	16.8	10	18.4
11	12.10	11	14.8	11	16.6	11	18.4	11	20.2
12	14.	12	16.	12	18.	12	20.	12	22.
13	15.2	13	17.4	13	19.6	13	21.8	13	23.10
14	16.4	14	18.8	14	21.	14	23.4	14	25.8
15	17.6	15	20.	15	22.6	15	25.	15	27.6
16	18.8	16	21.4	16	24.	16	26.8	16	29.4
17	19.10	17	22.8	17	25.6	17	28.4	17	31.2
18	21.	18	24.	18	27.	18	30.	18	33.
19	22.2	19	25.4	19	28.6	19	31.8	19	34.10
20	23.4	20	26.8	20	30.	20	33.4	20	36.8
21	24.6	21	28.	21	31.6	21	35.	21	38.6
22	25.8	22	29.4	22	33.	22	36.8	22	40.4
23	26.10	23	30.8	23	34.6	23	38.4	23	42.2
24	28.	24	32.	24	36.	24	40.	24	44.
25	29.2	25	33.4	25	37.6	25	41.8	25	45.10
26	30.4	26	34.8	26	39.	26	43.4	26	47.8
27	31.6	27	36.	27	40.6	27	45.	27	49.6
28	32.8	28	37.4	28	42.	28	46.8	28	51.4
29	33.10	29	38.8	29	43.6	29	48.4	29	53.2
30	35.	30	40.	30	45.	30	50.	30	55.

SCANTLING MEASURE

2½ x 5		2½ x 6		2½ x 7		2½ x 8		2½ x 9	
Length		Length		Length		Length		Length	
1	1.1	1	1.3	1	1.6	1	1.8	1	1.11
2	2.1	2	2.6	2	2.11	2	3.4	2	3.9
3	3.1	3	3.9	3	4.5	3	5.	3	5.8
4	4.2	4	5.	4	5.10	4	6.8	4	7.6
5	5.3	5	6.3	5	7.4	5	8.4	5	9.5
6	6.3	6	7.6	6	8.9	6	10.	6	11.3
7	7.4	7	8.9	7	10.3	7	11.8	7	13.2
8	8.4	8	10.	8	11.8	8	13.4	8	15.
9	9.5	9	11.3	9	13.2	9	15.	9	16.11
10	10.5	10	12.6	10	14.7	10	16.8	10	18.9
11	11.6	11	13.9	11	16.1	11	18.4	11	20.8
12	12.6	12	15.	12	17.6	12	20.	12	22.6
13	13.7	13	16.3	13	19.	13	21.8	13	24.5
14	14.7	14	17.6	14	20.5	14	23.4	14	26.3
15	15.8	15	18.9	15	21.11	15	25.	15	28.2
16	16.8	16	20.	16	23.4	16	26.8	16	30.
17	17.9	17	21.3	17	24.10	17	28.4	17	31.11
18	18.9	18	22.6	18	26.3	18	30.	18	33.9
19	19.10	19	23.9	19	27.9	19	31.8	19	35.8
20	20.10	20	25.	20	29.2	20	33.4	20	37.6
21	21.11	21	26.3	21	30.8	21	35.	21	39.5
22	22.11	22	27.6	22	32.1	22	36.8	22	41.3
23	24.	23	28.9	23	33.7	23	38.4	23	43.2
24	25.	24	30.	24	35.	24	40.	24	45.
25	26.1	25	31.3	25	36.6	25	41.8	25	46.11
26	27.1	26	32.6	26	37.11	26	43.4	26	48.9
27	28.2	27	33.9	27	39.5	27	45.	27	50.8
28	29.2	28	35.	28	40.10	28	46.8	28	52.6
29	30.3	29	36.3	29	42.4	29	48.4	29	54.5
30	31.3	30	37.6	30	43.9	30	50.	30	56.3

SCANTLING MEASURE

2½ x 10		2½ x 11		2½ x 12		3 x 3		3 x 4	
Length		Length		Length		Length		Length	
1	2.1	1	2.4	1	2.6	1	0.9	1	1.
2	4.2	2	4.7	2	5.	2	1.6	2	2.
3	6.3	3	6.11	3	7.6	3	2.3	3	3.
4	8.4	4	9.2	4	10.	4	3.	4	4.
5	10.5	5	11.6	5	12.6	5	3.9	5	5.
6	12.6	6	13.9	6	15.	6	4.6	6	6.
7	14.7	7	16.1	7	17.6	7	5.3	7	7.
8	16.8	8	18.4	8	20.	8	6.	8	8.
9	18.9	9	20.8	9	22.6	9	6.9	9	9.
10	20.10	10	22.11	10	25.	10	7.6	10	10.
11	22.11	11	25.3	11	27.6	11	8.3	11	11.
12	25.	12	27.6	12	30.	12	9.	12	12.
13	27.1	13	29.10	13	32.0	13	0.0	13	13
14	29.2	14	32.1	14	35.	14	10.6	14	14.
15	31.3	15	34.4	15	37.6	15	11.3	15	15.
16	33.4	16	36.8	16	40.	16	12.	16	16.
17	35.5	17	39.	17	42.6	17	12.9	17	17.
18	37.6	18	41.3	18	45.	18	13.6	18	18.
19	39.7	19	43.7	19	47.6	19	14.3	19	19.
20	41.8	20	45.10	20	50.	20	15.	20	20.
21	43.9	21	48.2	21	52.6	21	15.9	21	21.
22	45.10	22	50.5	22	55.	22	16.6	22	22.
23	47.11	23	52.9	23	57.6	23	17.3	23	23.
24	50.	24	55.	24	60.	24	18.	24	24.
25	52.1	25	57.4	25	62.6	25	18.9	25	25.
26	54.2	26	59.7	26	65.	26	19.6	26	26.
27	56.3	27	61.11	27	67.6	27	20.3	27	27.
28	58.4	28	64.2	28	70.	28	21.	28	28.
29	60.5	29	66.2	29	72.6	29	21.9	29	29.
30	62.6	30	68.9	30	75.	30	22.6	30	30.

SCANTLING MEASURE

3 x 5		3 x 6		3 x 7		3 x 8		3 x 9	
Length 1	1.3	Length 1	1.6	Length 1	1.9	1	2.	Length 1	2.3
2	2.6	2	3.	2	3.6	2	4.	2	4.6
3	3.9	3	4.6	3	5.3	3	6.	3	6.9
4	5.	4	6.	4	7.	4	8.	4	9.
5	6.3	5	7.6	5	8.9	5	10.	5	11.3
6	7.6	6	9.	6	10.6	6	12.	6	13.6
7	8.9	7	10.6	7	12.3	7	14.	7	15.9
8	10.	8	12.	8	14.	8	16.	8	18.
9	11.3	9	13.6	9	15.9	9	18.	9	20.3
10	12.6	10	15.	10	17.6	10	20.	10	22.6
11	13.9	11	16.6	11	19.3	11	22.	11	24.9
12	15.	12	18.	12	21.	12	24.	12	27.
13	16.3	13	19.6	13	22.9	13	26.	13	29.3
14	17.6	14	21.	14	24.6	14	28.	14	31.6
15	18.9	15	22.6	15	26.3	15	30.	15	33.9
16	20.	16	24.	16	28.	16	32.	16	36.
17	21.3	17	25.6	17	29.9	17	34.	17	38.3
18	22.6	18	27.	18	31.6	18	36.	18	40.6
19	23.9	19	28.6	19	33.3	19	38.	19	42.9
20	25.	20	30.	20	35.	20	40.	20	45.
21	26.3	21	31.6	21	36.9	21	42.	21	47.3
22	27.6	22	33.	22	38.6	22	44.	22	49.6
23	28.9	23	34.6	23	40.3	23	46.	23	51.9
24	30.	24	36.	24	42.	24	48.	24	54.
25	31.3	25	37.6	25	43.9	25	50.	25	56.3
26	32.6	26	39.	26	45.6	26	52.	26	58.6
27	33.9	27	40.6	27	47.3	27	54.	27	60.9
28	35.	28	42.	28	49.	28	56.	28	63.
29	36.3	29	43.6	29	50.9	29	58.	29	65.3
30	37.6	30	45.	30	52.6	30	60.	30	67.6

SCANTLING MEASURE

Length	3 x 10	3 x 11	3 x 12	4 x 4	5 x 4
1	2.6	2.9	3.	1.4	1.8
2	5.	5.6	6.	2.8	3.4
3	7.6	8.3	9.	4.	5.
4	10.	11.	12.	5.4	6.8
5	12.6	13.9	15.	6.8	8.4
6	15.	16.6	18.	8.	10.
7	17.6	19.3	21.	9.4	11.8
8	20.	22.	24.	10.8	13.4
9	22.6	24.9	27.	12.	15.
10	25.	27.6	30.	13.4	16.8
11	27.6	30.3	33.	14.8	18.4
12	30.	23.	36.	16.	20.
13	32.6	35.9	39.	17.4	21.8
14	35.	38.6	42.	18.8	23.4
15	37.6	41.3	45.	20.	25.
16	40.	44.	48.	21.4	26.8
17	42.6	46.9	51.	22.8	28.4
18	45.	49.6	54.	24.	30.
19	47.6	52.3	57.	25.4	31.8
20	50.	55.	60.	26.8	33.4
21	52.6	57.9	63.	28.	35.
22	55.	60.6	66.	29.4	36.8
23	57.6	63.3	69.	30.8	38.4
24	60.	66.	72.	32.	40.
25	62.6	68.9	75.	33.4	41.8
26	65.	71.6	78.	34.8	43.4
27	67.6	74.3	81.	36.	45.
28	70.	77.	84.	37.4	46.8
29	72.6	79.9	87.	38.8	48.4
30	75.	82.6	90.	40.	50.

SCANTLING MEASURE

	4 x 6		4 x 7		4 x 8		4 x 9		4 x 10
Length		Length		Length		Length		Length	
1	2.	1	2.4	1	2.8	1	3.	1	3.4
2	4.	2	4.8	2	5.4	2	6.	2	6.8
3	6.	3	7.	3	8	3	9	3	10.
4	8.	4	9.4	4	10.8	4	12.	4	13.4
5	10.	5	11.8	5	13.4	5	15.	5	16.8
6	12.	6	14.	6	16.	6	18.	6	20.
7	14.	7	16.4	7	18.8	7	21.	7	23.4
8	16.	8	18.8	8	21.4	8	24.	8	26.8
9	18.	9	21.	9	24.	9	27.	9	30.
10	20.	10	23.4	10	26.8	10	30.	10	33.4
11	22.	11	25.8	11	29.4	11	33.	11	36.8
12	24.	12	28.	12	32.	12	36.	12	40.
13	26.	13	30.4	13	34.8	13	39.	13	43.4
14	28.	14	32.8	14	37.4	14	42.	14	46.8
15	30.	15	35.	15	40.	15	45.	15	50.
16	32.	16	37.4	16	42.8	16	48.	16	53.4
17	34.	17	39.8	17	45.4	17	51.	17	56.8
18	36.	18	42.	18	48.	18	54.	18	60.
19	38.	19	44.4	19	50.8	19	57.	19	63.4
20	40.	20	46.8	20	53.4	20	60.	20	66.8
21	42.	21	49.	21	56.	21	63.	21	70.
22	44.	22	51.4	22	58.8	22	66.	22	73.4
23	46.	23	53.8	23	61.4	23	69.	23	76.8
24	48.	24	56.	24	64.	24	72.	24	80.
25	50.	25	58.4	25	66.8	25	75.	25	83.4
26	52.	26	60.8	26	69.4	26	78.	26	86.8
27	54.	27	63.	27	72.	27	81.	27	90.
28	56.	28	65.4	28	74.8	28	84.	28	93.4
29	58.	29	67.8	29	77.4	29	87.	29	96.8
30	60.	30	70.	30	80.	30	90.	30	100.

SCANTLING MEASURE

Length	4 x 11	Length	4 x 12	Length	5 x 5	Length	5 x 6	Length	5 x 7
1	3.8	1	4.	1	2.1	1	2.6	1	2.11
2	7.4	2	8.	2	4.2	2	5.	2	5.10
3	11.	3	12.	3	6.3	3	7.6	3	8.9
4	14.8	4	16.	4	8.4	4	10.	4	11.8
5	18.4	5	20.	5	10.5	5	12.6	5	14.7
6	22.	6	24.	6	12.6	6	15.	6	17.6
7	25.8	7	28.	7	14.7	7	17.6	7	20.5
8	29.5	8	32.	8	16.8	8	20.	8	23.4
9	33.	9	36.	9	18.9	9	22.6	9	26.3
10	36.8	10	40.	10	20.10	10	25.	10	29.2
11	40.4	11	44.	11	22.11	11	27.6	11	32.1
12	44.	12	48.	12	25.	12	30.	12	35.
13	47.8	13	52.	13	27.1	13	32.6	13	37.11
14	51.4	14	56.	14	29.2	14	35.	14	40.10
15	55.	15	60.	15	31.3	15	37.6	15	43.9
16	58.8	16	64.	16	33.4	16	40.	16	46.8
17	62.4	17	68.	17	35.5	17	42.6	17	49.7
18	66.	18	72.	18	37.6	18	45.	18	52.6
19	69.8	19	76.	19	39.7	19	47.6	19	55.5
20	73.4	20	80.	20	41.8	20	50.	20	58.4
21	77.	21	84.	21	43.9	21	52.6	21	61.3
22	80.8	22	88.	22	45.10	22	55.	22	64.2
23	84.4	23	92.	23	47.11	23	57.6	23	67.1
24	88.	24	96.	24	50.	24	60.	24	70.
25	91.8	25	100.	25	52.1	25	62.6	25	72.11
26	95.4	26	104.	26	54.2	26	65.	26	75.10
27	99.	27	108.	27	56.3	27	67.6	27	78.9
28	102.8	28	112.	28	58.4	28	70.	28	81.8
29	106.4	29	116.	29	60.5	29	72.6	29	84.7
30	110.	30	120.	30	62.6	30	75.	30	87.6

SCANTLING MEASURE

6 x 8		7 x 7		7 x 8		7 x 9		8 x 8	
Length		Length		Length		Length		Length	
1	4.	1	4.1	1	4.8	1	5.3	1	5.4
2	8.	2	8.2	2	9.4	2	10.6	2	10.8
3	12.	3	12.3	3	14.	3	15.9	3	16.
4	16.	4	16.4	4	18.8	4	21.	4	21.4
5	20.	5	20.5	5	23.4	5	26.3	5	26.8
6	24.	6	24.6	6	28.	6	31.6	6	32.
7	28.	7	28.7	7	32.8	7	36.9	7	37.4
8	32.	8	32.8	8	37.4	8	42.	8	42.8
9	36.	9	36.9	9	42.	9	47.3	9	48.
10	40.	10	40.10	10	46.8	10	52.6	10	53.4
11	44.	11	44.11	11	51.4	11	57.9	11	58.8
12	48.	12	49.	12	56.	12	63.	12	64.
13	52.	13	53.1	13	60.8	13	68.3	13	69.4
14	56.	14	57.2	14	65.4	14	73.6	14	74.8
15	60.	15	61.3	15	70.	15	78.9	15	80.
16	64.	16	65.4	16	74.8	16	84.	16	85.4
17	68.	17	69.5	17	79.4	17	89.3	17	90.8
18	72.	18	73.6	18	84.	18	94.6	18	96.
19	76.	19	77.7	19	88.8	19	99.9	19	101.4
20	80.	20	81.8	20	93.4	20	105.	20	106.8
21	84.	21	85.9	21	98.	21	110.3	21	112.
22	88.	22	89.10	22	102.8	22	115.6	22	117.4
23	92.	23	93.11	23	107.4	23	120.9	23	122.8
24	96.	24	98.	24	112.	24	126.	24	128.
25	100.	25	102.1	25	116.8	25	131.3	25	133.4
26	104.	26	106.2	26	121.4	26	136.6	26	138.8
27	108.	27	110.3	27	126.	27	141.9	27	144.
28	112.	28	114.4	28	130.8	28	147.	28	149.4
29	116.	29	118.5	29	135.4	29	152.3	29	154.8
30	120.	30	122.6	30	140.	30	157.6	30	160.

SCANTLING MEASURE

	8 x 9		8 x 10		9 x 9		9 x 10		9 x 11
1	6.	1	6.8	1	6.9	1	7.6	1	8.3
2	12.	2	13.4	2	13.6	2	15.	2	16.6
3	18.	3	20.	3	20.3	3	22.6	3	24.9
4	24.	4	26.8	4	27.	4	30.	4	33.
5	30.	5	33.4	5	33.9	5	37.6	5	41.3
6	36.	6	40.	6	40.6	6	45.	6	49.6
7	42.	7	46.8	7	47.3	7	52.6	7	57.9
8	48.	8	53.4	8	54.	8	60.	8	66.
9	54.	9	60.	9	60.9	9	67.6	9	74.3
10	60.	10	66.8	10	67.6	10	75.	10	82.6
11	66.	11	73.4	11	74.3	11	82.6	11	90.9
12	72.	12	80.	12	81.	12	90.	12	99.
13	78.	13	86.8	13	87.9	13	97.6	13	107.3
14	84.	14	93.4	14	94.6	14	105.	14	115.6
15	90.	15	100.	15	101.3	15	112.6	15	123.9
16	96.	16	106.8	16	108.	16	120.	16	132.
17	102.	17	113.4	17	114.9	17	127.6	17	140.3
18	108.	18	120.	18	121.6	18	135.	18	148.6
19	114.	19	126.8	19	128.3	19	142.6	19	156.9
20	120.	20	133.4	20	135.	20	150.	20	165.
21	126.	21	140.	21	141.9	21	157.6	21	173.3
22	132.	22	146.8	22	148.6	22	165.	22	181.6
23	138.	23	153.4	23	155.3	23	172.6	23	189.9
24	144.	24	160.	24	162.	24	180.	24	198.
25	150.	25	166.8	25	168.9	25	187.6	25	206.3
26	156.	26	173.4	26	175.6	26	195.	26	214.6
27	162.	27	180.	27	182.3	27	202.6	27	222.9
28	168.	28	186.8	28	189.	28	210.	28	231.
29	174.	29	193.4	29	195.9	29	217.6	29	239.3
30	180.	30	200.	30	202.6	30	225.	30	247.6

SCANTLING MEASURE

	10 x 10		10 x 11		10 x 12		11 x 11		11 x 12
Length 1	8.4	Length 1	9.2	Length 1	10.	Length 1	10.1	Length 1	11.
2	16.8	2	18.4	2	20.	2	20.2	2	22.
3	25.	3	27.6	3	30.	3	30.3	3	33.
4	33.4	4	36.8	4	40.	4	40.4	4	44.
5	41.8	5	45.10	5	50.	5	50.5	5	55.
6	50.	6	55.	6	60.	6	60.6	6	66.
7	58.4	7	64.2	7	70.	7	70.7	7	77.
8	66.8	8	73.4	8	80.	8	80.8	8	88.
9	75.	9	82.6	9	90.	9	90.9	9	99.
10	83.4	10	91.8	10	100.	10	100.10	10	110.
11	91.8	11	100.10	11	110.	11	110.11	11	121.
12	100.	12	110.	12	120.	12	121.	12	132.
13	108.4	13	119.2	13	130.	13	131.1	13	143.
14	116.8	14	128.4	14	140.	14	141.2	14	154.
15	125.	15	137.6	15	150.	15	151.3	15	165.
16	133.4	16	146.8	16	160.	16	161.4	16	176.
17	141.8	17	155.10	17	170.	17	171.5	17	187.
18	150.	18	165.	18	180.	18	181.6	18	198.
19	158.4	19	174.2	19	190.	19	191.7	19	209.
20	166.8	20	183.4	20	200.	20	201.8	20	220.
21	175.	21	192.6	21	210.	21	211.9	21	231.
22	183.4	22	201.8	22	220.	22	221.10	22	242.
23	191.8	23	210.10	23	230.	23	231.11	23	253.
24	200.	24	220.	24	240.	24	242.	24	264.
25	208.4	25	229.2	25	250.	25	252.1	25	275.
26	216.8	26	238.4	26	260.	26	262.2	26	286.
27	225.	27	247.6	27	270.	27	272.3	27	297.
28	233.4	28	256.8	28	280.	28	282.4	28	308.
29	241.8	29	265.10	29	290.	29	292.5	29	319.
30	250.	30	275.	30	300.	30	302.6	30	330.

CONDENSED SCANTLING TABLE

Showing the Number of Feet, B.M., Contained in a
Piece of Joist, Scantling or Timber,
of the sizes given

LENGTH IN FEET

Size in inches	12	14	16	18	20	22	24	26	28	30
2x4 ..	8	9	11	12	13	15	16	17	19	20
2x6 ..	12	14	16	18	20	22	24	26	28	30
2x8 ..	16	19	21	24	27	29	32	35	37	40
2x10..	20	23	27	30	33	37	40	43	47	50
2x12..	24	28	32	36	40	44	48	52	56	60
3x4 ..	12	14	16	18	20	22	24	26	28	30
3x6 ..	18	21	24	27	30	33	36	39	42	45
3x8 ..	24	28	32	36	40	44	48	52	56	60
3x10..	30	35	40	45	50	55	60	65	70	75
3x12..	36	42	48	54	60	66	72	78	84	90
4x4 ..	16	19	21	24	27	29	32	35	37	40
4x6 ..	24	28	32	36	40	44	48	52	56	60
6x6 ..	36	42	48	54	60	66	72	78	84	90
6x8 ..	48	56	64	72	80	88	96	104	112	120
8x8 ..	64	75	85	96	107	117	128	139	149	160
8x10..	80	93	107	120	133	147	160	173	187	200
10x10..	100	117	133	150	167	183	200	217	233	250
10x12..	120	140	160	180	200	220	240	260	280	300
12x12..	144	168	192	216	240	264	288	312	336	360

AN ADJUSTABLE SAW BUCK

Take two forked tree limbs, of good size (as shown by the cut), bore a two inch hole through from the under side at the proper angle, and you have a very convenient, adjustable and cheap saw buck. It always rests firmly upon the ground, while the upper end is a crotch to hold the wood; very convenient for cutting up stove wood, or for holding timber or lumber of any kind.

CULTIVATE BLACK WALNUT, the supply is fast being exhausted, while the demand for that kind of wood for furniture and other purposes is very great. Trees of good size grow in 10 to 12 years, and the lumber commands a very high price.

BOARD MEASURE

EXPLANATION

The length of any board will be found in feet at the top of the column, and the width in inches in the left hand column.

To find the number of feet, B. M., in any board, find the length at the top of the column and the width in the left hand column; trace the lines until they meet, and you will find the amount sought for. For example: On page 32, a board 10 feet long and 18 inches wide is shown to contain fifteen feet, board measure.

BRIEF REMARKS

Besides inch boards, plank and scantling are usually bought and sold by board measure; round, sawed or hewn timber is bought and sold by the cubic foot.

Pine and spruce spars, from 10 to 4½ inches in diameter, inclusive, are measured by taking the diameter, clear of bark, at one-third of their length at the large end.

Spars are usually purchased by the inch diameter; all under four inches are considered *poles*.

Boards are sold by the square foot surface, one inch in thickness.

The dimensions of a foot of board measure are 1 foot long, 1 foot high, and 1 inch thick.

BOARD MEASURE

LENGTH IN FEET

Inches wide	4	5	6	7	8	9	10
6	2.00	2.06	3.00	3.06	4.00	4.06	5.00
7	2.04	2.11	3.06	4.01	4.08	5.03	5.10
8	2.08	3.04	4.00	4.08	5.04	6.00	6.08
9	3.00	3.09	4.06	5.03	6.00	6.09	7.06
10	3.04	4.02	5.00	5.10	6.08	7.06	8.04
11	3.08	4.07	5.06	6.05	7.04	8.03	9.02
12	4.00	5.00	6.00	7.00	8.00	9.00	10.00
13	4.04	5.05	6.06	7.07	8.08	9.09	10.10
14	4.08	5.10	7.00	8.02	9.04	10.06	11.08
15	5.00	6.03	7.06	8.09	10.00	11.03	12.06
16	5.04	6.08	8.00	9.04	10.08	12.00	13.04
17	5.08	7.01	8.06	9.11	11.04	12.09	14.02
18	6.00	7.06	9.00	10.06	12.00	13.06	15.00
19	6.04	7.11	9.06	11.01	12.08	14.03	15.10
20	6.08	8.04	10.00	11.08	13.04	15.00	16.08
21	7.00	8.09	10.06	12.03	14.00	15.09	17.06
22	7.04	9.02	11.00	12.10	14.08	16.06	18.04
23	7.08	9.07	11.06	13.05	15.04	17.03	19.02
24	8.00	10.00	12.00	14.00	16.00	18.00	20.00
25	8.04	10.05	12.06	14.07	16.08	18.09	20.10
26	8.08	10.10	13.00	15.02	17.04	19.06	21.08
27	9.00	11.03	13.06	15.09	18.00	20.03	22.06
28	9.04	11.08	14.00	16.04	18.08	21.00	23.04
29	9.08	12.01	14.06	16.11	19.04	21.09	24.02
30	10.00	12.06	15.00	17.06	20.00	22.06	25.00

₊ The width is in the margin—length at the head.

BOARD MEASURE

LENGTH IN FEET

Inches wide	11	12	13	14	15	16	17
3	2.09	3.00	3.03	3.06	3.09	4.00	4.03
4	3.08	4.00	4.04	4.08	5.00	5.04	5.08
5	4.07	5.00	5.05	5.10	6.03	6.08	7.01
6	5.06	6.00	6.06	7.00	7.06	8.00	8.06
7	6.05	7.00	7.07	8.02	8.09	9.04	9.11
8	7.04	8.00	8.08	9.04	10.00	10.08	11.04
9	8.03	9.00	9.09	10.06	11.03	12.00	12.09
10	9.02	10.00	10.10	11.08	12.06	13.04	14.02
11	10.01	11.00	11.11	12.10	13.09	14.08	15.07
12	11.00	12.00	13.00	14.00	15.00	16.00	17.10
13	11.11	13.00	14.01	15.02	16.03	17.04	18.05
14	12.10	14.00	15.02	16.04	17.06	18.08	19.00
15	13.09	15.00	16.03	17.06	18.09	20.00	21.03
16	14.08	16.00	17.04	18.08	20.00	21.04	22.08
17	15.07	17.00	18.05	19.10	21.03	22.08	24.01
18	16.06	18.00	19.06	21.00	22.06	24.00	25.06
19	17.05	19.00	20.07	22.02	23.09	25.04	26.11
20	18.04	20.00	21.08	23.04	25.00	26.08	28.04
21	19.03	21.00	22.09	24.06	26.03	28.00	29.09
22	20.02	22.00	23.10	25.08	27.06	29.04	31.02
23	21.01	23.00	24.11	26.10	28.09	30.08	32.07
24	22.00	24.00	26.00	28.00	30.00	32.00	34.00
25	22.11	25.00	27.01	29.02	31.03	33.04	35.05
26	23.10	26.00	28.02	30.04	32.06	34.08	36.10
27	24.09	27.00	29.03	31.06	33.09	36.00	38.03
28	25.08	28.00	30.04	32.08	35.00	37.04	39.08
29	26.07	29.00	31.05	33.10	36.03	38.08	41.01
30	27.06	30.00	32.06	35.00	37.06	40.00	42.06

*** The width is in the margin—length at the head.

BOARD MEASURE

LENGTH IN FEET

Inches wide	18	19	20	21	22	23	24
3	4.06	4.09	5.00	5.03	5.06	5.09	6.00
4	6.00	6.04	6.08	7.00	7.04	7.08	8.00
5	7.03	7.11	8.04	8.09	9.02	9.07	10.00
6	9.00	9.06	10.00	10.06	11.00	11.06	12.00
7	10.06	11.01	11.08	12.03	12.10	13.05	14.00
8	12.00	12.08	13.04	14.00	14.08	15.04	16.00
9	13.06	14.03	15.00	15.09	16.06	17.03	18.00
10	15.00	15.10	16.08	17.06	18.04	19.02	20.00
11	16.06	17.05	18.04	19.03	20.02	21.01	22.00
12	18.00	19.00	20.00	21.00	22.00	23.00	24.00
13	19.06	20.07	21.08	22.09	23.10	24.11	26.00
14	21.00	22.02	23.04	24.06	25.08	26.10	28.00
15	22.06	23.09	25.00	26.03	27.06	28.09	30.00
16	24.00	25.04	26.08	28.00	29.04	30.08	32.00
17	25.06	26.11	28.04	29.09	31.02	32.07	34.00
18	27.00	28.06	30.00	31.06	33.00	34.06	36.00
19	28.06	30.01	31.08	33.03	34.10	36.05	38.00
20	30.00	31.08	33.04	35.00	36.08	38.04	40.00
21	31.06	33.03	35.00	36.09	38.06	40.03	42.00
22	33.00	34.10	36.08	38.06	40.04	42.02	44.00
23	34.06	36.05	38.04	40.03	42.02	44.01	46.00
24	36.00	38.00	40.00	42.00	44.00	46.00	48.00
25	37.06	39.07	41.08	43.09	45.10	47.11	50.00
26	39.00	41.02	43.04	45.06	47.08	49.10	52.00
27	40.06	42.09	45.00	47.03	49.06	51.09	54.00
28	42.00	44.04	46.08	49.00	51.04	53.08	56.00
29	43.06	45.11	48.04	50.09	53.02	55.07	58.00
30	45.00	47.06	50.00	52.06	55.00	57.06	60.00

*** The width is in the margin—length at the head.

PLANK MEASURE

Board measure is the basis of plank measure; that is, a plank *two* inches thick and 13 feet long and 10 inches wide, contains, evidently, twice as many square feet as if only one inch thick.

EXPLANATION

The following tables show at one view, the number of feet, board measure, contained in any ship, or other plank, from 24 to 52 feet in length, and from 1¾ inches in thickness to 4, varying from ¼ to ½ an inch, and from 10 inches to 28 in width.

The length of any plank will be found in the left hand column of the table, and the width and thickness at the head of the page.

To find the number of feet which any plank will give, take the length in the left hand column of the table, and the width and thickness at the top of the page—trace the two lines until they meet, and you will have the amount.

For Example: A plank 47 feet in length, 2½ inches thick, by 23 inches in width, will give 225 feet, the required sum. If the plank exceeds in length any provision which is made in these tables, its contents would be shown by taking twice what is given for half its length; and for a lesser length, half what is shown for twice its length. In all cases, in these computations, the smaller fractions of a foot are omitted, while the larger ones are reckoned a foot; this is sufficiently correct for all practical purposes.

PLANK MEASURE

L. Ft.	$1\frac{3}{4}$ x 10	$1\frac{3}{4}$ x 11	$1\frac{3}{4}$ x 12	$1\frac{3}{4}$ x 13	$1\frac{3}{4}$ x 14	$1\frac{3}{4}$ x 15	$1\frac{3}{4}$ x 16	$1\frac{3}{4}$ x 17	$1\frac{3}{4}$ x 18
24	35	39	42	45	49	52	56	59	63
25	36	40	44	47	51	55	59	62	66
26	38	42	45	49	53	57	61	64	68
27	39	43	47	51	55	59	63	67	71
28	41	45	49	53	57	61	65	69	73
29	42	47	51	55	59	63	68	72	76
30	43	48	52	57	61	65	70	74	79
31	45	50	54	59	63	68	72	77	81
32	47	51	56	61	65	70	75	79	84
33	48	53	58	63	67	72	77	82	87
34	49	55	59	64	69	74	79	84	89
35	51	56	61	66	71	76	82	87	92
36	52	58	63	68	73	79	84	89	94
37	54	59	65	70	75	81	86	92	97
38	55	61	66	72	77	83	89	94	100
39	57	63	68	74	80	85	91	97	102
40	58	64	70	76	82	87	93	99	105
41	60	67	72	78	84	90	96	102	108
42	61	69	73	80	86	92	98	104	110
43	62	71	75	82	88	94	100	108	113
44	64	73	77	83	90	96	103	109	115
45	65	74	79	85	92	98	105	112	118
46	67	76	80	87	94	101	107	114	121
47	68	78	82	89	96	103	110	117	123
48	70	79	84	91	98	105	112	119	126
49	71	80	86	93	100	107	114	121	129
50	73	82	88	94	102	109	117	124	131
51	74	83	89	97	104	112	119	126	134
52	76	84	91	99	106	114	121	129	136

PLANK MEASURE

L. Ft.	$1\frac{3}{4}$ x 19	$1\frac{3}{4}$ x 20	$1\frac{3}{4}$ x 21	$1\frac{3}{4}$ x 22	$1\frac{3}{4}$ x 23	$1\frac{3}{4}$ x 24	$1\frac{3}{4}$ x 25	$1\frac{3}{4}$ x 26	$1\frac{3}{4}$ x 27
24	66	70	74	77	81	84	88	91	94
25	69	73	77	80	84	87	91	95	98
26	72	76	80	83	87	91	95	99	102
27	74	79	83	87	91	94	98	102	106
28	77	82	86	90	94	98	102	106	110
29	80	85	89	93	97	101	106	110	114
30	83	87	92	96	101	105	109	114	118
31	85	90	95	99	104	108	113	118	122
32	88	93	98	103	107	112	117	121	126
33	91	96	101	106	111	115	120	125	130
34	93	99	104	109	114	119	124	129	134
35	95	102	107	112	117	122	128	133	138
36	98	105	110	115	121	126	131	136	142
37	101	108	114	119	124	129	135	140	146
38	104	111	117	122	127	133	138	144	150
39	107	114	120	125	131	136	142	148	154
40	109	117	123	128	134	140	146	152	158
41	112	120	126	132	137	143	149	155	162
42	115	122	129	135	141	147	153	159	166
43	118	125	132	138	144	150	156	163	170
44	121	128	135	141	147	154	160	167	174
45	123	131	138	144	151	157	164	171	178
46	126	134	141	148	154	161	167	174	180
47	129	137	144	151	158	164	171	178	184
48	132	140	147	154	161	168	174	182	188
49	134	143	150	157	164	171	178	186	192
50	136	146	153	160	168	175	182	190	196
51	139	149	156	164	171	178	185	193	200
52	141	152	159	167	174	182	189	197	204

PLANK MEASURE

L. Ft.	2 x 12	2 x 13	2 x 14	2 x 15	2 x 16	2 x 17	2 x 18	2 x 19	2 x 20
24	48	52	56	60	64	68	72	76	80
25	50	54	58	62	67	71	75	79	83
26	52	56	61	65	69	74	78	82	87
27	54	59	63	67	72	76	81	85	90
28	56	61	65	70	75	79	84	89	93
29	58	63	68	72	77	82	87	92	97
30	60	65	70	75	80	85	90	95	100
31	62	67	72	77	83	88	93	98	103
32	64	69	75	80	85	91	96	101	107
33	66	71	77	82	88	93	99	104	110
34	68	74	79	85	91	96	102	108	113
35	70	76	82	87	93	99	105	111	117
36	72	78	84	90	96	102	108	114	120
37	74	80	86	92	99	105	111	117	123
38	76	82	89	95	101	108	114	120	127
39	78	84	91	97	104	110	117	123	130
40	80	87	93	100	107	113	120	127	133
41	82	88	96	102	109	116	123	130	137
42	84	91	98	105	112	119	126	133	140
43	86	93	100	107	115	122	129	136	143
44	88	95	103	110	117	125	132	139	147
45	90	97	105	112	120	127	135	142	150
46	92	99	107	115	123	130	138	146	153
47	94	102	110	117	125	133	141	149	157
48	96	104	112	120	128	136	144	152	160
49	98	106	114	122	131	139	147	155	163
50	100	108	117	125	133	142	150	158	167
51	102	110	119	127	136	144	153	161	170
52	104	112	121	130	139	147	156	165	173

PLANK MEASURE

L. Ft.	2 x 21	2 x 22	2 x 23	2 x 24	2 x 25	2 x 26	2 x 27	2 x 28	2¼ x 10
24	84	88	92	96	100	104	108	112	45
25	87	92	96	100	104	108	112	117	47
26	91	95	100	104	108	113	117	121	49
27	94	99	103	108	112	117	121	126	51
28	98	103	107	112	117	121	126	131	52
29	101	106	111	116	121	126	130	139	54
30	105	110	115	120	125	130	135	140	56
31	108	114	119	124	129	134	139	145	58
32	112	117	123	128	133	139	144	149	60
33	115	121	126	132	137	143	148	154	62
34	119	125	130	136	141	147	153	159	64
35	122	128	134	140	146	152	157	163	66
36	126	132	138	144	150	156	162	168	67
37	129	136	142	148	154	160	166	173	69
38	133	139	146	152	158	165	171	177	71
39	136	143	149	156	162	169	175	182	73
40	140	147	153	160	167	173	180	187	75
41	143	150	157	164	171	178	184	191	77
42	147	154	161	168	175	182	189	196	79
43	150	158	165	172	179	186	193	201	81
44	154	161	169	176	183	191	198	205	83
45	157	165	172	180	187	195	202	210	84
46	160	169	176	184	192	199	207	215	86
47	164	172	180	188	196	204	211	219	88
48	168	176	184	192	200	208	216	224	90
49	171	180	188	196	204	212	220	229	92
50	175	183	192	200	208	217	225	233	94
51	178	187	195	204	213	221	229	238	96
52	182	190	199	208	217	225	234	243	98

PLANK MEASURE

L. Ft.	$2\frac{1}{4}$ x 11	$2\frac{1}{4}$ x 12	$2\frac{1}{4}$ x 13	$2\frac{1}{4}$ x 14	$2\frac{1}{4}$ x 15	$2\frac{1}{4}$ x 16	$2\frac{1}{4}$ x 17	$2\frac{1}{4}$ x 18	$2\frac{1}{4}$ x 19
24	49	54	58	64	68	72	77	80	86
25	52	56	61	66	70	75	80	84	89
26	54	59	63	68	73	78	83	88	93
27	56	61	66	71	76	81	86	91	96
28	58	63	68	73	79	84	89	94	100
29	60	65	71	76	82	87	92	98	103
30	62	68	73	79	84	90	96	101	107
31	64	70	76	81	87	93	99	105	110
32	66	72	78	84	90	96	101	108	114
33	68	74	80	86	93	99	104	111	118
34	70	77	83	89	96	102	109	115	121
35	72	79	85	92	98	105	112	118	125
36	74	81	88	94	101	108	115	121	128
37	76	83	90	97	104	111	118	125	132
38	78	86	92	100	107	114	121	128	135
39	80	88	95	102	110	117	124	131	139
40	82	90	97	105	112	120	128	135	142
41	85	92	100	107	115	123	131	138	146
42	87	95	102	110	118	126	134	142	150
43	89	97	105	113	122	129	137	145	153
44	91	99	107	115	125	132	140	148	157
45	93	102	110	118	127	135	144	152	160
46	95	104	112	121	130	138	147	155	164
47	97	106	115	123	133	141	150	159	167
48	99	108	117	126	136	144	153	162	171
49	101	111	119	128	139	147	156	165	175
50	103	113	122	131	141	150	159	169	178
51	105	115	124	134	144	153	163	172	182
52	107	117	127	136	146	156	166	175	185

LUMBER AND LOG BOOK   39

PLANK MEASURE

L. Ft.	2¼ x 20	2¼ x 21	2¼ x 22	2¼ x 23	2¼ x 24	2¼ x 25	2¼ x 26	2¼ x 27	2½ x 12
24	90	95	99	104	108	113	117	121	60
25	94	98	103	108	112	117	122	127	62
26	97	102	107	112	117	122	127	132	65
27	101	106	111	116	121	127	132	137	67
28	105	110	115	121	126	131	136	142	70
29	109	114	119	125	130	136	141	147	72
30	112	118	123	129	135	141	146	152	75
31	116	122	127	134	139	145	151	157	77
32	120	126	131	138	144	150	156	162	80
33	124	130	135	142	148	155	161	167	82
34	127	134	140	147	153	159	166	172	85
35	131	138	144	151	157	164	171	177	87
36	135	142	148	155	162	169	175	182	90
37	139	146	152	160	166	173	180	187	92
38	143	150	156	164	171	178	185	192	95
39	146	154	160	168	175	183	190	197	97
40	150	158	164	172	180	187	195	202	100
41	154	162	168	177	184	192	200	207	102
42	158	166	172	181	189	197	205	213	105
43	162	170	177	185	193	201	210	218	107
44	165	174	181	190	198	206	215	223	110
45	169	178	185	194	202	211	220	228	112
46	173	182	189	198	207	215	224	232	115
47	176	186	193	203	211	220	229	238	117
48	180	189	197	207	216	225	234	243	120
49	183	193	201	211	220	229	239	248	122
50	187	197	205	215	225	234	244	253	125
51	191	201	210	220	229	239	249	258	127
52	195	205	214	224	234	244	254	263	130

PLANK MEASURE

L. Ft.	$2\frac{1}{2}$ x 13	$2\frac{1}{2}$ x 14	$2\frac{1}{2}$ x 15	$2\frac{1}{2}$ x 16	$2\frac{1}{2}$ x 17	$2\frac{1}{2}$ x 18	$2\frac{1}{2}$ x 19	$2\frac{1}{2}$ x 20	$2\frac{1}{2}$ x 21
24	65	70	75	80	85	90	95	100	105
25	68	73	78	83	89	94	99	104	109
26	70	76	81	87	92	97	103	108	114
27	73	79	84	90	96	101	107	112	118
28	76	82	87	93	99	105	111	117	122
29	79	85	91	97	103	109	115	121	127
30	81	87	94	100	106	112	119	125	131
31	84	90	97	103	110	116	123	129	136
32	86	93	100	107	113	120	127	133	140
33	89	96	103	110	117	124	131	137	144
34	92	99	106	113	120	127	135	142	149
35	95	102	109	117	124	131	139	146	153
36	97	105	113	120	127	135	143	150	157
37	100	108	116	123	131	139	147	154	162
38	103	111	119	127	135	142	150	158	166
39	106	114	122	130	138	146	154	162	171
40	108	117	125	133	142	150	158	168	175
41	111	120	128	137	145	154	162	171	179
42	114	122	131	140	149	157	166	175	184
43	116	125	134	143	152	161	170	179	188
44	119	128	137	147	156	165	174	183	192
45	122	131	140	150	159	169	178	187	197
46	125	134	144	153	163	172	182	192	201
47	127	137	147	157	166	176	186	196	206
48	130	140	150	160	170	180	190	200	210
49	133	143	153	163	174	184	194	204	215
50	135	146	156	167	177	187	198	208	219
51	138	149	159	170	181	191	202	212	223
52	141	152	162	173	185	195	206	216	227

PLANK MEASURE

L. Ft.	2½ x 22	2½ x 23	2½ x 24	2½ x 25	2½ x 26	2½ x 27	2½ x 28	3 x 12	3 x 13
24	110	115	120	125	130	135	140	72	78
25	115	120	125	130	135	141	146	75	81
26	119	125	130	135	141	146	152	78	84
27	124	129	135	141	146	152	157	81	88
28	128	134	140	146	152	158	163	84	91
29	133	139	145	151	157	163	169	87	94
30	137	144	150	156	163	169	175	90	98
31	142	148	155	161	168	175	181	93	101
32	147	153	160	167	173	180	187	96	104
33	151	158	165	172	179	186	192	99	107
34	156	163	170	177	184	191	198	102	111
35	160	168	175	182	190	197	204	105	114
36	165	172	180	187	195	203	210	108	117
37	171	177	185	193	200	208	216	111	120
38	174	182	190	198	206	214	222	114	123
39	179	187	195	203	211	220	227	117	127
40	183	192	200	208	217	225	233	120	130
41	188	196	205	214	222	231	239	123	133
42	192	201	210	219	228	237	245	126	136
43	197	206	215	224	233	242	251	129	140
44	202	211	220	229	238	248	256	132	143
45	206	216	225	234	244	253	262	135	146
46	211	220	230	240	249	259	268	138	149
47	215	225	235	245	254	265	274	141	152
48	220	230	240	250	260	270	280	144	156
49	225	235	245	255	265	276	286	147	159
50	229	240	250	260	271	282	292	150	162
51	234	244	255	266	276	289	298	153	165
52	238	249	260	270	282	293	303	156	169

PLANK MEASURE

L. Ft.	3 x 14	3 x 15	3 x 16	3 x 17	3 x 18	3 x 19	3 x 20	3 x 21	3 x 22
24	84	90	96	102	108	114	120	126	132
25	87	94	100	106	112	119	125	131	138
26	91	97	104	110	117	123	130	136	143
27	94	101	108	115	121	128	135	142	149
28	98	105	112	119	126	133	140	147	154
29	101	109	116	123	130	138	145	152	160
30	105	112	120	127	135	142	150	157	165
31	108	116	124	132	139	147	155	163	171
32	112	120	128	136	144	151	160	168	176
33	115	124	132	140	149	156	165	173	182
34	119	127	136	144	153	161	170	178	187
35	122	131	140	149	157	166	175	184	193
36	126	135	144	153	162	170	180	189	198
37	129	139	148	157	166	175	185	194	204
38	133	142	152	161	171	180	190	199	209
39	136	146	156	166	175	185	195	204	215
40	140	150	160	170	180	189	200	210	220
41	143	154	164	174	184	194	205	215	226
42	147	157	168	178	189	199	210	220	231
43	150	161	172	183	193	204	215	225	236
44	154	165	176	187	198	208	220	231	242
45	157	169	180	191	202	213	225	236	247
46	161	172	184	195	207	218	230	241	253
47	164	176	188	200	211	223	235	247	258
48	168	180	192	204	216	227	240	252	264
49	171	184	196	208	220	232	245	257	269
50	175	187	200	212	225	237	250	262	275
51	178	191	204	217	229	242	255	268	280
52	181	195	208	221	234	246	260	274	286

PLANK MEASURE

L. Ft.	3 x 23	3 x 24	3 x 25	3 x 26	3 x 27	3 x 28	3 x 29	3 x 30	3½ x 15
24	138	144	150	156	162	168	174	180	105
25	144	150	156	162	169	175	181	187	109
26	149	156	162	169	175	182	188	195	114
27	155	162	169	175	182	189	196	202	118
28	161	168	175	182	189	196	203	210	122
29	167	174	181	188	196	203	210	217	127
30	172	180	187	195	202	210	217	225	131
31	178	186	194	201	209	217	225	232	136
32	184	192	200	208	216	224	232	240	140
33	190	198	206	214	223	231	239	247	144
34	195	204	212	221	225	238	246	255	149
35	201	210	219	227	236	245	254	262	153
36	207	216	225	234	243	252	261	270	157
37	213	222	231	240	250	259	268	277	162
38	218	228	237	247	256	266	275	285	166
39	224	234	244	253	263	273	283	292	171
40	230	240	250	260	270	280	290	300	175
41	236	246	256	266	277	287	297	307	179
42	241	252	262	273	283	294	304	315	184
43	247	258	269	279	290	301	312	322	188
44	253	264	275	286	297	308	319	330	192
45	259	270	281	292	304	315	326	337	197
46	264	276	287	299	310	322	333	345	201
47	270	282	294	305	317	329	341	352	206
48	276	288	300	312	324	336	348	360	210
49	282	294	306	318	331	343	355	367	214
50	287	300	312	325	337	350	362	375	219
51	293	306	319	331	344	357	370	382	223
52	299	312	325	338	351	364	377	390	227

PLANK MEASURE

L. Ft.	3½ x 16	3½ x 17	3½ x 18	3½ x 19	3½ x 20	3½ x 21	3½ x 22	3½ x 23	3½ x 24
24	112	119	126	133	140	147	154	161	168
25	117	124	131	139	146	153	160	168	175
26	121	129	136	144	152	159	167	174	182
27	126	134	142	150	157	165	173	181	189
28	131	139	147	155	163	172	180	188	196
29	135	144	152	161	169	178	186	195	203
30	140	149	157	167	175	184	192	201	210
31	145	154	163	172	181	190	199	208	217
32	149	159	168	177	187	196	205	215	224
33	154	164	173	183	192	202	212	221	231
34	159	169	178	188	195	208	218	228	238
35	163	174	184	194	204	214	225	235	245
36	168	178	189	200	210	221	231	241	252
37	173	183	194	205	216	227	237	248	259
38	177	188	199	211	222	233	244	255	266
39	182	193	205	216	227	239	250	262	273
40	186	198	210	222	233	245	257	268	280
41	191	203	215	227	239	251	263	275	287
42	196	208	220	233	245	257	269	282	294
43	200	213	226	238	251	263	276	288	301
44	205	218	231	244	257	269	282	295	308
45	210	223	236	249	262	275	289	302	315
46	214	228	242	255	268	281	295	309	322
47	219	233	247	260	274	287	302	315	329
48	224	238	252	266	280	294	308	322	336
49	228	243	257	271	286	300	314	329	343
50	233	248	262	277	292	306	321	336	350
51	238	253	268	282	296	312	327	342	357
52	242	258	273	288	303	318	334	348	364

PLANK MEASURE

L. Ft.	3½ x 25	3½ x 26	3½ x 27	3½ x 28	3½ x 29	3½ x 30	4 x 15	4 x 16	4 x 17
24	175	182	189	196	203	210	120	128	136
25	182	190	197	204	211	219	125	133	142
26	190	197	205	212	220	227	130	139	147
27	197	205	213	220	228	236	135	144	153
28	204	212	220	229	237	245	140	149	159
29	211	220	228	237	245	254	145	155	164
30	219	227	236	245	254	262	150	160	170
31	226	235	244	253	262	271	155	165	176
32	233	243	252	261	271	280	160	171	181
33	241	250	260	269	279	289	165	176	187
34	248	258	268	278	287	297	170	181	193
35	255	265	276	286	296	306	175	187	198
36	262	273	283	294	304	315	180	192	204
37	269	281	291	302	313	324	185	197	210
38	277	288	299	310	321	332	190	203	215
39	284	296	307	318	330	341	195	208	221
40	292	303	315	327	338	350	200	213	227
41	299	311	323	335	346	359	205	219	232
42	306	318	331	343	355	367	210	224	238
43	313	326	339	351	363	376	215	229	244
44	320	333	346	359	372	385	220	235	249
45	328	341	354	367	381	394	225	240	255
46	335	349	362	376	389	402	230	245	261
47	342	356	370	384	397	411	235	251	266
48	350	364	378	392	406	420	240	256	272
49	356	372	386	400	414	429	245	261	278
50	365	379	394	408	423	437	250	267	283
51	372	387	402	416	431	446	255	272	289
52	379	394	409	424	440	454	260	278	294

PLANK MEASURE

L. Ft.	4 x 18	4 x 19	4 x 20	4 x 21	4 x 22	4 x 23	4 x 24	4 x 25	4 x 26
24	144	152	160	168	176	184	192	200	208
25	150	158	167	175	183	192	200	208	217
26	156	165	173	182	191	199	208	217	225
27	162	171	180	189	198	207	216	225	234
28	168	177	187	196	205	215	224	233	243
29	174	184	193	203	213	222	232	242	251
30	180	190	200	210	220	230	240	250	260
31	186	196	207	217	227	238	248	258	269
32	192	203	213	224	235	245	256	267	277
33	198	209	220	231	242	253	264	275	286
34	204	215	227	238	249	261	272	283	295
35	210	222	233	245	257	268	280	291	303
36	216	228	240	252	264	276	288	300	312
37	222	234	247	259	271	284	296	308	321
38	228	241	253	266	279	291	304	317	329
39	234	247	260	273	286	299	312	325	338
40	240	253	267	280	293	307	320	333	347
41	246	260	273	287	301	314	328	342	355
42	252	266	280	294	308	322	336	350	364
43	258	272	287	301	315	330	344	358	373
44	264	279	293	308	323	337	352	367	381
45	270	285	300	315	330	345	360	375	390
46	276	291	307	322	337	353	368	383	399
47	282	298	313	329	345	360	376	392	407
48	288	304	320	336	352	368	384	400	416
49	294	310	327	343	359	376	392	408	425
50	300	317	333	350	367	383	400	417	433
51	306	323	340	357	374	391	408	425	442
52	312	329	347	364	381	399	416	434	450

SQUARE TIMBER

EXPLANATION

The length of any stick of hewed or sawed timber will be found in the left hand column of the table; the side dimensions at the head of the page, and the cubical, or solid contents, may be found directly under the side dimensions, and at the right of the length. Thus, a stick of timber (page 51), measuring 10 by 12 inches, side dimensions, and 30 feet in length, contains 25 cubic feet of timber. So, also, a stick 20 by 22 inches, side dimensions, and 35 feet long, contains 107 cubic feet.

If a piece of timber should exceed, in length, any provision made in these tables, its contents may be found by taking twice what is shown for half its length, etc. Thus, a stick of timber 64 feet long would contain twice what is shown in the table for one 32 feet long, and so on.

When a stick of timber is larger at one end than at the other, the mean diameter, or square, must be sought for, and its contents computed from it.

In these computations, the decimal parts of a foot are omitted, when half or less than half a foot; and when more, they are reckoned as a whole foot. This will be sufficiently correct for all ordinary purposes.

NOTE.—Hewed timber for framing buildings, and for building bridges, docks, ships, &c., is sold by the solid cubic foot; and the contents of each stick, when measured by the lumberman, is marked on the butt with a broad-axe in Roman capital letters. For example, a stick containing nineteen feet is marked XIX., one twenty feet, XX., and so on. A cubic foot is a measurement one foot long by a foot thick each way, or the equivalent thereof; hence a stick of timber a foot square will count one cubic foot to each foot of its running length.

CUBICAL CONTENTS OF SQUARE TIMBER

L. Ft.	6 x 6	6 x 7	6 x 8	6 x 9	6 x 10	6 x 11	6 x 12
20	5.00	5.83	6.66	7.50	8.33	9.77	10.00
21	5.25	6.12	7.00	7.87	8.75	9.62	10.50
22	5.50	6.42	7.33	8.25	9.16	10.08	11.00
23	5.75	6.70	7.66	8.62	9.58	10.54	11.50
24	6.00	7.00	8.00	9.00	10.00	11.00	12.00
25	6.25	7.29	8.33	9.37	10.42	11.46	12.50
26	6.50	7.58	8.66	9.75	10.83	12.02	13.00
27	6.75	7.87	9.00	10.12	11.25	12.37	13.50
28	7.00	8.16	9.33	10.50	11.66	12.83	14.00
29	7.25	8.45	9.66	10.87	12.08	13.29	14.50
30	7.50	8.75	10.00	11.25	12.50	13.75	15.00
31	7.75	9.04	10.33	11.62	12.92	14.21	15.50
32	8.00	9.33	10.66	12.00	13.33	14.66	16.00
33	8.25	9.62	11.00	12.37	13.75	15.12	16.50
34	8.50	9.91	11.33	12.75	14.17	15.59	17.00
35	8.75	10.20	11.66	13.12	14.58	16.04	17.50
36	9.00	10.50	12.00	13.50	15.00	16.50	18.00
37	9.25	10.79	12.33	13.87	15.42	16.96	18.50
38	9.50	11.08	12.66	14.25	15.83	17.41	19.00
39	9.75	11.37	13.00	14.62	16.25	17.87	19.50
40	10.00	11.66	13.33	15.00	16.66	18.33	20.00
41	10.25	11.95	13.66	15.37	17.08	18.79	20.50
42	10.50	12.25	14.00	15.75	17.50	19.25	21.00
43	10.75	12.54	14.33	16.12	17.92	19.71	21.50
44	11.00	12.83	14.66	16.50	18.33	20.16	22.00
45	11.25	13.12	15.00	16.87	18.75	20.62	22.50
46	11.50	13.41	15.33	17.25	19.17	21.08	23.00
47	11.75	13.70	15.66	17.62	19.58	21.54	23.50
48	12.00	14.00	16.00	18.00	20.00	22.00	24.00

CUBICAL CONTENTS OF SQUARE TIMBER

L. Ft.	7 x 7	7 x 8	7 x 9	7 x 10	7 x 11	7 x 12	8 x 8
20	6.80	7.77	8.75	9.72	10.69	11.66	8.88
21	7.14	8.16	9.18	10.20	11.23	12.25	9.33
22	7.48	8.55	9.62	10.69	11.76	12.83	9.77
23	7.82	8.94	10.06	11.18	12.29	13.41	10.22
24	8.16	9.33	10.50	11.66	12.83	14.00	10.66
25	8.50	9.72	10.93	12.15	13.37	14.58	11.11
26	8.84	10.11	11.37	12.64	13.90	15.16	11.55
27	9.18	10.50	11.81	13.12	14.44	15.75	12.00
28	9.52	10.88	12.25	13.61	14.94	16.33	12.44
29	9.87	11.27	12.68	14.09	15.50	16.91	12.88
30	10.20	11.66	13.12	14.58	16.04	17.50	13.33
31	10.54	12.05	13.56	15.07	16.57	17.58	13.77
32	10.89	12.44	14.00	15.55	17.11	18.66	14.22
33	11.23	12.83	14.43	16.04	17.64	19.25	14.66
34	11.57	13.22	14.87	16.52	18.18	19.83	15.11
35	11.91	13.61	15.31	17.01	18.71	20.41	15.55
36	12.25	14.00	15.75	17.50	19.25	21.00	16.00
37	12.59	14.39	16.18	17.98	19.78	21.58	16.44
38	12.93	14.77	16.62	18.47	20.32	22.16	16.88
39	13.27	15.16	17.06	18.96	20.85	22.75	17.33
40	13.61	15.55	17.50	19.44	21.39	23.33	17.77
41	13.95	15.94	17.93	19.93	21.87	23.91	18.22
42	14.29	16.33	18.37	20.41	22.46	24.50	18.66
43	14.63	16.72	18.81	20.90	22.99	25.08	19.11
44	14.97	17.11	19.25	21.38	23.52	25.66	19.55
45	15.31	17.50	19.68	21.87	24.06	26.25	20.00
46	15.65	17.89	20.12	22.36	24.59	26.83	20.44
47	16.00	18.27	20.56	22.84	25.13	27.41	20.88
48	16.33	18.66	21.00	23.33	25.66	28.00	21.33

CUBICAL CONTENTS OF SQUARE TIMBER

L. ft.	8 x 9	8 x 10	8 x 11	8 x 12	9 x 9	9 x 10	9 x 11	9 x 12
20	10.00	11.11	12.22	13.33	11.25	12.50	13.75	15.00
21	10.50	11.66	12.83	14.00	11.81	13.12	14.44	15.75
22	11.00	12.22	13.44	14.66	12.37	13.75	15.12	16.50
23	11.50	12.77	14.05	15.33	12.93	14.37	15.81	17.25
24	12.00	13.33	14.66	16.00	13.50	15.00	16.50	18.00
25	12.50	13.88	15.27	16.66	14.06	15.62	17.18	18.75
26	13.00	14.44	15.88	17.33	14.62	16.25	17.87	19.50
27	13.50	15.00	16.50	18.00	15.18	16.87	18.56	20.25
28	14.00	15.55	17.11	18.66	15.75	17.50	19.25	21.00
29	14.50	16.11	17.72	19.33	16.31	18.12	19.93	21.75
30	15.00	16.66	18.33	20.00	16.87	18.75	20.62	22.50
31	15.50	17.22	18.94	20.66	17.43	19.37	21.31	23.25
32	16.00	17.77	19.55	21.33	18.00	20.00	22.00	24.00
33	16.50	18.33	20.16	22.00	18.56	20.62	22.68	24.75
34	17.00	18.88	20.77	22.66	19.12	21.25	23.37	25.50
35	17.50	19.44	21.39	23.33	19.68	21.87	24.06	26.25
36	18.00	20.00	22.00	24.00	20.25	22.50	24.75	27.00
37	18.50	20.55	22.61	24.66	20.81	23.12	25.43	27.75
38	19.00	21.11	23.22	25.33	21.37	23.75	26.12	28.50
39	19.50	21.66	23.83	26.00	21.93	24.37	26.81	29.25
40	20.00	22.22	24.44	26.66	22.50	25.00	27.50	30.00
41	20.50	22.77	25.05	27.33	23.06	25.62	28.18	30.75
42	21.00	23.33	25.66	28.00	23.62	26.25	28.87	31.50
43	21.50	23.88	26.27	28.66	24.18	26.87	29.56	32.25
44	22.00	24.44	26.88	29.33	24.75	27.50	30.25	33.00
45	22.50	25.00	27.50	30.00	25.31	28.12	30.93	33.75
46	23.00	25.55	28.11	30.66	25.87	28.75	31.62	34.50
47	23.50	26.11	28.71	31.33	26.43	29.37	32.31	35.25
48	24.00	26.66	29.33	32.00	27.00	30.00	33.00	36.00

CUBICAL CONTENTS OF SQUARE TIMBER

L. ft.	10 x 10	10 x 11	10 x 12	10 x 13	11 x 11	11 x 12	11 x 13	11 x 14	12 x 12
20	14	15	17	18	17	18	20	21	20
21	15	16	17	19	18	19	21	22	21
22	15	17	18	20	18	20	22	23	22
23	16	18	19	21	19	21	23	25	23
24	17	18	20	22	20	22	24	26	24
25	17	19	21	23	21	23	25	27	25
26	18	20	22	23	22	24	26	28	26
27	19	21	22	24	23	25	27	29	27
28	19	21	23	25	23	26	28	30	28
29	20	22	24	26	24	27	29	31	29
30	21	23	25	27	25	28	30	32	30
31	21	24	26	28	26	28	31	33	31
32	22	24	27	29	27	29	32	34	32
33	23	25	28	30	28	30	33	35	33
34	24	26	28	31	29	31	34	36	34
35	24	27	29	32	29	32	35	37	35
36	25	27	30	32	30	33	36	38	36
37	26	28	31	33	31	34	37	40	37
38	26	29	32	34	32	35	38	41	38
39	27	30	32	35	33	36	39	42	39
40	28	31	33	36	34	37	40	43	40
41	28	31	34	37	34	38	41	44	41
42	29	32	35	38	35	38	42	45	42
43	30	33	36	39	36	39	43	46	43
44	31	34	37	40	37	40	44	47	44
45	31	34	37	41	38	41	45	48	45
46	32	35	38	41	39	42	46	49	46
47	33	36	39	42	39	43	47	50	47
48	33	37	40	43	40	44	48	51	48

CUBICAL CONTENTS OF SQUARE TIMBER

L. ft.	12 x 13	12 x 14	12 x 15	13 x 13	13 x 14	13 x 15	13 x 16	14 x 14	14 x 15
20	22	23	25	23	25	27	29	27	29
21	23	24	26	25	27	28	30	29	31
22	24	26	27	26	28	30	32	30	32
23	25	27	29	27	29	31	33	31	34
24	26	28	30	28	30	32	35	33	35
25	27	29	31	29	32	34	36	34	36
26	28	30	32	30	33	35	38	35	38
27	29	31	34	32	34	37	39	37	39
28	30	32	35	33	35	38	40	38	41
29	31	34	36	34	37	39	42	39	42
30	32	35	37	35	38	41	43	41	44
31	34	36	39	36	39	42	45	42	45
32	35	37	40	38	40	43	46	44	47
33	36	38	41	39	42	45	48	45	48
34	37	40	42	40	43	46	49	46	50
35	38	41	44	41	44	47	51	48	51
36	39	42	45	42	45	49	52	49	52
37	40	43	46	43	47	50	53	50	54
38	41	44	47	45	48	51	55	52	55
39	42	45	49	46	49	53	56	53	57
40	43	47	50	47	51	54	58	54	58
41	44	48	51	48	52	55	59	56	60
42	45	49	52	49	53	57	61	57	61
43	46	50	54	50	54	58	62	58	63
44	48	51	55	52	56	60	64	60	64
45	49	52	56	53	57	61	65	61	66
46	50	54	57	54	58	62	66	63	67
47	51	55	58	55	59	64	68	64	69
48	52	56	60	56	61	65	69	65	70

CUBICAL CONTENTS OF SQUARE TIMBER

L. ft.	14 x 16	14 x 17	15 x 15	15 x 16	15 x 17	15 x 18	16 x 16	16 x 17	16 x 18
20	31	33	31	33	35	37	36	38	40
21	33	35	33	35	37	39	37	40	42
22	34	36	34	37	39	41	39	42	44
23	36	38	36	38	41	43	41	43	46
24	37	40	37	40	42	45	43	45	48
25	39	41	39	42	44	47	44	47	50
26	40	43	41	43	46	49	46	49	52
27	42	45	42	45	48	51	48	51	54
28	44	46	44	47	50	52	50	53	56
29	45	48	45	48	51	54	52	55	58
30	47	50	47	50	53	56	53	57	60
31	48	51	48	52	55	58	55	59	62
32	50	53	50	53	57	60	57	60	64
33	51	55	52	55	58	62	59	62	66
34	53	56	53	57	60	64	60	64	68
35	54	58	55	58	62	66	62	66	70
36	56	59	56	60	64	67	64	68	72
37	58	61	58	62	65	69	66	70	74
38	59	63	59	63	67	71	68	72	76
39	61	64	61	65	69	73	69	74	78
40	62	66	62	67	71	75	71	76	80
41	64	68	64	68	73	77	73	77	82
42	65	69	66	70	74	79	75	79	84
43	67	71	67	72	76	81	76	81	86
44	68	73	69	73	78	82	78	83	88
45	70	74	70	75	80	84	80	85	90
46	72	76	72	77	81	86	82	87	92
47	73	78	73	78	83	88	84	89	94
48	75	79	74	80	85	90	85	91	96

CUBICAL CONTENTS OF SQUARE TIMBER

L. ft.	16 x 19	17 x 17	17 x 18	17 x 19	17 x 20	18 x 18	18 x 19	18 x 20	18 x 21
20	42	40	42	45	47	45	47	50	52
21	44	42	45	47	50	47	50	52	55
22	46	44	47	49	52	49	52	55	58
23	49	46	49	52	54	52	55	57	60
24	51	48	51	54	57	54	57	60	63
25	53	50	53	56	59	56	59	62	66
26	55	52	55	58	61	58	62	65	68
27	57	54	57	61	64	61	64	67	71
28	59	56	59	63	66	63	66	70	73
29	61	58	62	65	68	65	69	72	76
30	63	60	64	67	71	67	71	75	79
31	65	62	66	70	73	70	74	77	81
32	68	64	68	72	76	72	76	80	84
33	70	66	70	74	78	74	78	82	87
34	72	68	72	76	80	76	81	85	89
35	74	70	74	79	83	79	83	87	92
36	76	72	76	81	85	81	85	90	94
37	78	74	79	83	87	83	88	92	97
38	80	76	81	85	90	85	90	95	100
39	82	78	83	88	92	88	93	97	102
40	84	80	85	90	94	90	95	100	105
41	87	82	87	92	97	92	97	102	108
42	89	84	89	94	99	94	100	105	110
43	91	86	91	97	101	97	102	107	113
44	93	88	93	99	103	99	104	110	115
45	95	90	96	101	106	101	107	112	118
46	97	92	98	104	109	104	109	115	121
47	99	94	100	105	111	106	112	117	123
48	101	96	102	108	113	108	114	120	126

CUBICAL CONTENTS OF SQUARE TIMBER

L. ft.	19 x 19	19 x 20	19 x 21	19 x 22	20 x 20	20 x 21	20 x 22	20 x 23	21 x 21
20	50	53	55	58	56	58	61	63	61
21	53	55	58	61	58	61	64	67	64
22	55	58	61	64	61	64	67	70	67
23	58	61	64	67	64	67	70	73	70
24	60	63	66	70	67	70	73	76	73
25	63	66	69	73	69	73	76	78	77
26	65	69	72	76	72	76	79	83	80
27	68	71	75	78	75	79	82	86	83
28	70	74	78	81	78	82	86	89	86
29	73	76	80	84	81	85	89	92	89
30	75	79	83	87	83	87	92	95	92
31	78	82	86	90	86	90	95	98	95
32	80	84	89	93	89	93	98	101	98
33	83	87	91	96	92	96	101	104	101
34	85	90	94	99	94	99	104	108	104
35	88	92	97	102	97	102	107	111	107
36	90	95	100	104	100	105	110	115	110
37	93	98	103	107	103	106	113	118	113
38	95	100	105	110	106	111	116	121	116
39	98	103	108	113	108	114	119	124	119
40	100	106	111	116	111	117	122	127	122
41	103	108	114	119	114	120	125	130	126
42	105	111	110	122	117	122	128	134	120
43	108	113	119	125	119	125	132	137	132
44	110	116	122	128	122	128	135	140	135
45	113	119	125	131	125	131	138	143	138
46	115	121	128	134	128	134	140	146	141
47	118	124	130	136	131	137	144	150	144
48	120	127	133	139	133	140	147	153	147

CUBICAL CONTENTS OF SQUARE TIMBER

L. ft.	21 x 22	21 x 23	21 x 24	22 x 22	22 x 23	22 x 24	22 x 25	23 x 23	23 x 24
20	64	67	70	67	70	73	76	73	76
21	67	70	73	70	73	77	80	77	80
22	71	73	77	73	77	80	84	80	84
23	74	76	80	77	80	84	87	84	88
24	77	80	84	80	84	88	91	88	92
25	80	83	87	83	87	91	95	91	95
26	83	87	91	87	91	95	99	95	99
27	87	90	94	90	94	99	103	99	103
28	90	93	98	93	98	102	106	102	107
29	93	97	101	97	101	106	110	106	111
30	96	100	105	100	105	110	114	110	115
31	99	103	108	103	108	113	118	113	118
32	103	107	112	107	112	117	122	117	122
33	106	110	115	110	115	121	126	121	126
34	109	114	119	114	119	124	129	124	130
35	112	117	122	117	122	128	133	128	134
36	115	120	126	121	126	132	137	132	138
37	119	124	129	124	130	135	141	135	141
38	122	127	133	127	133	139	145	139	145
39	125	130	136	131	137	143	148	143	149
40	128	134	140	134	140	146	152	146	153
41	131	137	143	137	144	150	156	150	157
42	134	140	147	141	147	154	160	154	161
43	138	144	150	144	151	157	164	157	164
44	141	147	154	147	154	161	168	161	168
45	144	150	157	151	158	165	171	165	172
46	148	154	161	154	161	168	175	168	176
47	151	157	164	157	163	172	179	172	180
48	154	161	168	161	165	176	183	176	184

CUBICAL CONTENTS OF SQUARE TIMBER

L. ft.	23 x 25	24 x 24	24 x 25	24 x 26	24 x 27	25 x 25	25 x 26	25 x 27	25 x 28
20	79	80	83	86	90	86	90	93	97
21	83	84	87	91	94	91	94	98	102
22	87	88	91	95	99	95	99	103	106
23	91	92	95	99	103	99	103	107	111
24	95	96	100	104	108	104	108	112	116
25	99	100	104	108	112	108	112	117	121
26	103	104	108	112	117	112	117	121	126
27	107	108	112	117	121	117	121	126	131
28	111	112	116	121	126	121	126	131	136
29	115	116	120	125	130	125	130	135	140
30	119	120	125	130	135	130	135	140	145
31	123	124	129	134	139	134	139	145	150
32	127	128	133	138	144	138	144	150	155
33	131	132	137	143	148	143	148	154	160
34	135	136	141	147	153	147	153	159	165
35	139	140	145	151	157	151	157	164	170
36	143	144	150	156	162	156	162	168	175
37	147	148	154	160	166	160	167	173	179
38	151	152	158	164	171	164	171	178	184
39	155	156	162	169	175	169	176	182	189
40	159	160	166	173	180	173	180	187	194
41	163	164	170	177	184	177	185	192	199
42	167	168	175	182	189	182	189	196	204
43	171	172	179	186	193	186	194	201	209
44	175	176	183	190	198	190	198	206	213
45	179	180	187	195	202	195	203	210	218
46	183	184	191	199	207	199	207	215	223
47	187	188	195	203	211	204	212	220	228
48	191	192	200	208	216	208	216	225	233

NUMBER OF PIECES

REQUIRED FOR 1,000 FEET BOARD MEASURE

(Fractions omitted.)

Lgth.	12ft.	14ft.	16ft.	18 ft.	20 ft.	22 ft.	24 ft.
SIZE	PCS.	PCS.	PCS.	PCS.	PCS.	PCS.	PCS.
2 x 4	125	108	94	84	75	69	63
1x8	1000	1008	1002	1008	1000	1012	1008
2 x 6	84	72	63	56	50	46	42
1x12	1008	1008	1008	1008	1000	1012	1008
2 x 8	63	54	47	42	38	35	32
4x4	1008	1008	1002	1008	1013	1026	1024
2 x 10	50	43	38	34	30	28	25
4x5	1000	1003	1013	1020	1000	1026	1000
2 x 12	42	36	32	28	25	23	21
3x8	1008	1008	1024	1008	1000	1012	1008
4 x 8	32	27	24	21	19	18	16
2x16	1024	1008	1024	1008	1013	1056	1024
3 x 10	34	29	25	23	20	19	17
2x15	1020	1015	1000	1035	1000	1045	1020
3 x 12	28	24	21	19	17	16	14
6x6	1008	1008	1008	1026	1020	1056	1008
2 x 14	36	31	27	24	22	20	18
4x7	1008	1012	1008	1008	1026	1026	1008
3 x 14	24	21	18	16	15	13	12
6x7	1008	1029	1008	1008	1050	1001	1008

CUBIC MEASUREMENT

EXPLANATION.—The length of any log, in feet, will be found in the left hand column of the table, and the *average diameter*, in inches, may be found at the head of the page. Thus, a log 19 inches diameter, and 38 feet long, contains 43 ft., and six-twelfths, cubic measurement.

REMARKS

These tables have been computed from the following:
RULE.—Add together the two extreme diameters, and divide by two for the mean diameter. Subtract one-third for the side of the square the log will make when hewn. Square the side thus obtained, and multiply the product by the length of the log in feet, and divide the last product by 144 (or by twelve twice), the quotient will be the cubical contents in feet, and twelfths of a foot.

This rule, after much consultation with both buyer and seller of lumber, is, I believe, more nearly the truth than any other that can be made, and this is conceded by all *sellers* of lumber with whom I have conversed; besides, it has attained almost universal use in practice. This rule does not give quite so much as the square *inscribed* in a circle equal to the diameter of the log, but as trees never grow perfectly round nor straight, some waste will be experienced, and allowance ought justly to be made to the purchaser, from the mathematical accuracy of inscribing a square in a circle. The average diameter may also be taken in sections of 15 feet, or by the rule above, as the parties may agree.

As the above rule corresponds with universal practice, these tables may, with propriety, be regarded as the *Standard Tables* for reducing round timber to square, cubical measurement.

EXAMPLE.—Suppose the adjoining diagram to represent a log, whose extreme diameters are 18 and 24 inches, and 45 feet long—how many cubic feet does it contain?

Length 45 ft

OPERATION.

18+24=42; and 42÷2=21 inches, average diameter.
⅓ of 21=7. Then, 21−7=14; and 14^2=196;
196×45÷144=61 ft. 3 in. *Ans.*

NOTE.—The diameter multiplied by .7071, gives the side of the square any round log will make when squared.

ROUND TIMBER REDUCED TO SQUARE TIMBER

CUBIC MEASUREMENT.

L. ft.	Av. Dia. 12	Av. Dia. 13	Av. Dia. 14	Av. Dia. 15	Av. Dia. 16	Av. Dia. 17	Av. Dia. 18	Av. Dia. 19
25	11.1	14.1	15.1	17.4	19.10	22.6	25.0	28.0
26	11.6	14.8	15.8	18.1	20.8	23.0	26.0	29.1
27	12.	15.2	16.3	18.9	21.5	23.11	27.0	30.2
28	12.5	15.9	16.10	19.5	22.3	24.10	28.0	31.4
29	12.10	16.4	17.5	20.2	23.0	25.9	29.0	32.5
30	13.3	16.11	18.	20.10	23.10	26.8	30.0	33.6
31	13.8	17.5	18.7	21.6	24.7	27.7	31.0	34.8
32	14.2	18.0	19.2	22.3	25.5	28.6	32.0	35.9
33	14.7	18.7	19.9	22.11	26.2	29.5	33.0	36.11
34	15.	19.2	20.4	23.7	27.0	30.4	34.0	38.0
35	15.6	19.8	20.11	24.4	27.9	31.3	35.0	40.2
36	15.11	20.3	21.6	25.0	28.7	32.2	36.0	41.3
37	16.4	20.10	22.1	25.8	29.5	33.1	37.0	42.5
38	16.10	21.5	22.8	26.5	30.2	34.0	38.0	43.6
39	17.4	21.11	23.4	27.1	31.10	34.11	39.0	44.7
40	17.9	22.6	24.0	27.9	31.9	35.10	40.0	45.9
41	18.3	23.1	24.7	28.6	32.7	36.9	41 0	46 10
42	18.8	23.8	25.2	29.2	33.4	37.8	42.0	48.0
43	19.1	24.2	25.9	29.10	34.2	38.7	43.0	49.1
44	19.7	24.9	26.4	30.7	34.11	39.6	44.0	50.3
45	20.	25.4	27.0	31.3	35.9	40.5	45.0	51.4
46	20.5	25.11	27.7	31.11	36.6	41.4	46.0	52.6
47	20.11	26.5	28.2	32.8	37.4	42.3	47.0	53.7
48	21.4	27.	28.9	33.4	38.1	43.2	48.0	54.9
49	21.9	27.7	29.4	34.0	38.11	44.1	49.0	55.10
50	22.2	28.2	30.0	34.8	39.8	45	50.0	56.0

NOTE.—The diameter multiplied by .7071 gives the side of the square any round log will make when squared.

ROUND TIMBER REDUCED TO SQUARE
TIMBER

CUBIC MEASUREMENT

L. ft.	Av. Dia. 20	Av. Dia. 21	Av. Dia. 22	Av. Dia. 23	Av. Dia. 24	Av. Dia. 25	Av. Dia. 26	Av. Dia. 27
25	31.8	34.0	39.1	41.9	44.5	50.2	53.2	56.3
26	32.11	35.5	40.8	43.5	46.2	52.2	55.4	58.6
27	34.2	36.9	42.2	45.1	48.0	54.2	57.5	60.9
28	35.5	38.1	43.9	46.9	49.8	56.2	59.7	63.0
29	36.8	39.6	45.4	48.5	50.4	58.2	61.8	65 3
30	38.0	40.10	46.11	50.1	52.1	60.3	63.10	67.6
31	39.3	42.2	48.5	51.9	54.0	62.3	65.11	69.9
32	40.6	43.7	50.0	53.5	55 9	64.3	68.1	72.0
33	41.9	44.11	51.7	55.1	57.6	66.3	70.2	74.3
34	43.0	46.3	53.2	56.9	59.3	68.3	72.4	76.6
35	44.4	47.8	54.8	58.5	61 1	70.3	74.5	78.9
36	45.7	49.0	56.3	60.1	62.9	72.3	76.7	81.0
37	46.10	50.4	57.10	61.9	64 6	74.3	78.8	83.3
38	47.1	51.9	59.5	63.5	66.3	74.3	80.10	85.6
39	49.4	53.1	60.11	64.1	68.0	76.3	82.11	87.9
40	50.8	54.5	62.6	66.9	69.9	78.3	85.1	90.0
41	51.11	55.10	64.1	68.5	71.6	80.3	87.2	92.3
42	53.2	57.2	65.8	70.1	73.3	82.4	89.4	94.6
43	54.5	58.6	67.2	71.9	75.0	84.4	91.5	96.9
44	55.8	59.11	68.9	73.5	77.9	86.4	93.8	99.0
45	56.11	61.3	70.4	75.1	79.6	88.4	95.8	101.3
46	58.3	62.7	71.11	76.9	81.3	90.4	97.10	103.6
47	59.6	64.0	73.5	78.5	83.0	92.4	99.11	105.9
48	60.9	65.4	75.0	80.1	84.9	94.4	102.1	108.0
49	62.0	66.8	76 7	81.9	86.6	96.4	104.3	110.3
50	63.3	68.1	78.2	83.5	88.3	98.4	106.4	112.6

ROUND TIMBER REDUCED TO SQUARE TIMBER

CUBIC MEASUREMENT

L ft.	Av. Dia. 28	Av. Dia. 29	Av. Dia. 30	Av. Dia. 31	Av. Dia. 32	Av. Dia. 33	Av. Dia. 34
25	62.8	66.8	69.5	73.0	73.9	84.0	88.0
26	65.2	69.4	72.3	75.11	81.11	87.5	91.5
27	67.8	72.0	75.0	78.10	85.1	90.9	95.0
28	70.2	74.8	77.9	81.9	88.3	94.1	98.5
29	72.8	77.4	80.7	84.8	91.5	97.6	102.0
30	75.3	80.0	83.4	87.7	94.7	100.0	105.6
31	77.9	82.8	86.1	90.6	97.9	104.2	109.0
32	80.3	85.4	88.11	93.5	100.11	107.7	112.6
33	82.9	88.0	91.8	96.4	104.1	111.0	116.0
34	85.3	90.8	94.5	99.3	107.3	114.3	119.6
35	87.9	93.4	97.3	102.2	110.5	117.8	123.1
36	90.3	96.0	100.0	105.1	113.7	121.0	126.7
37	92.9	98.8	102.9	108.0	116.9	124.4	130.1
38	95.3	101.4	105.7	110.11	119.11	127.9	133.7
39	97.9	104.0	108.4	113.10	123.1	131.1	137.1
40	100.3	106.8	111.1	116.9	126.3	134.5	140.8
41	102.9	109.4	113.11	119.8	129.5	138.0	144.2
42	105.4	112.0	116.8	122.7	132.7	141.2	147.8
43	107.10	114.8	119.5	125.6	135.9	144.6	151.2
44	110.4	117.4	122.3	128.5	138.11	148.0	154.8
45	112.10	120.0	125.0	131.4	142.1	151.3	158.2
46	115.4	122.8	127.9	134.3	145.3	154.7	161.9
47	117.10	125.4	130.7	137.2	148.5	158.0	165.3
48	120.4	128.0	133.4	140.1	151.7	161.4	168.9
49	123.10	130.8	136.1	143.0	154.9	164.8	172.3
50	125.4	133.4	138.11	145.11	157.10	168.1	175.9

ROUND TIMBER REDUCED TO SQUARE TIMBER

CUBIC MEASUREMENT

L. ft.	Av. Dia. 35	Av. Dia. 36	Av. Dia. 37	Av. Dia. 38	Av. Dia. 39	Av. Dia. 40	Av. Dia. 41
25	95.11	100.0	108.6	112.11	121.11	131.4	141.0
26	99.9	104.0	112.10	117.5	126.10	136.7	146.8
27	103.7	108.0	117.2	121.11	131.8	141.10	152.4
28	107.5	112.0	121.6	126.5	136.7	147.1	157.1
29	111.3	116.0	125.10	130.11	141.5	152.4	163.7
30	115.1	120.0	130.3	135.6	146.4	157.7	169.3
31	118.11	124.0	134.7	140.0	151.2	162.10	174.1
32	122.9	128.0	138.11	144.6	156.1	168.1	180.6
33	126.7	132.0	143.3	149.0	160.11	173.4	186.2
34	130.5	136.0	147.7	153.6	165.10	178.7	191.9
35	134.3	140.0	151.11	158.1	170.8	183.10	197.5
36	138.1	144.0	156.3	162.7	175.7	189.1	203.1
37	141.11	148.0	160.7	167.1	180.5	194.4	208.8
38	145.9	152.0	164.11	171.7	185.4	199.7	214.4
39	149.7	156.0	169.3	176.1	190.2	204.10	220.0
40	153.5	160.0	173.7	180.8	195.1	210.1	225.7
41	157.3	164.0	177.11	185.2	199.11	215.4	231.2
42	161.1	168.0	182.4	189.8	204.10	220.7	236.1
43	164.11	172.0	186.8	194.2	209.8	225.10	242.7
44	168.9	176.0	191.0	198.8	214.7	231.1	248.2
45	172.7	180.0	195.4	203.2	219.5	236.4	253.1
46	176.5	184.0	199.8	207.9	224.4	241.7	259.6
47	180.3	188.0	204.4	212.3	229.2	246.10	265.1
48	184.1	192.0	212.8	216.9	234.1	252.1	270.9
49	187.11	196.0	217.0	221.3	239.0	257.4	276.5
50	191.9	200.0	221.4	225.9	243.10	262.7	282.0

PROPERTIES OF WOODS

NAMES	Specific Gravity Water 1,000	Average Wt. of a Cu. ft in lbs.	Cubic Feet in a Ton	Comparative		
				Stiffness	Strength	Resistance
Eng. Oak.......	934	56	38½	100	100	100
Amer. Oak......	672	42	53	114	96	64
Beech..........	852	43	45	77	103	138
Sycamore.......	604	38	59	59	81	111
Chestnut........	630	38	59	67	89	118
Ash............	845	52	43	89	119	160
Elm............	673	42	53	78	82	86
Mahog. Sp......	800	50	45	73	67	61
Walnut.........	671	42	53	49	74	111
Poplar..........	383	54	66	44	50	57
Cedar..........	561	33	68	23	62	106
Amer. Spruce....	561	34	66	72	80	102
Yel. Pine.......	461	28	80	95	99	103
Pitch Pine......	600	41	54½	73	82	92
Larch..........	550	31	72	79	103	134

WERE IT not for dry rot, ships would last, on the average, about 30 years As it is, their average duration, when built of ordinary timber, is seven, eight and nine years.

TO MARK TOOLS.—Warm them slightly and rub the steel with wax, or hard tallow, till a film gathers. Then write your name on the wax with a sharp point, cutting through to the steel. A little nitric acid poured on the marking will bite in the letters. Then wipe the acid and wax off with a hot, soft rag.

SHOWING THE CUBICAL CONTENTS OF SPARS AND OTHER ROUND TIMBER

EXPLANATION AND REMARKS

The length of any spar, or log, will be found in feet in the left hand column of the table, and the *average diameter* in *inches*, may be seen at the top of the page—advancing in size 1 inch, from 10 to 38 inches

To find the cubic or solid contents which any spar or log will give, take the length in feet in the left hand column of the table, and the diameter in inches at the top of the page—trace the two lines until they meet, and you will have the amount sought for. Thus, a spar, or log, whose average diameter is 28 inches, and 36 feet in length, contains, according to our showing, 154 cubic feet; and one 34 inches diameter, and 28 feet long, 178. If a spar should exceed in length any provision made in these tables (as will often be the case), its contents may be found by taking twice what is shown for half its length. Thus a log 68 feet long, and 26 inches diameter, would contain twice what is shown in the table for one 34 feet long; *i.e.*, 252 feet. In these computations, the decimal parts of a foot are omitted, when half, or less than half, and when more, they are reckoned as a whole foot. This will be sufficiently correct for all ordinary purposes.

NOTE.—In computing the solidity of spars or logs in rafts, for charging toll, about 10 per cent. from these estimates should be deducted for the sudden taper of many logs, as also for the inequality of the diameters of the same log, and the protuberances of the bark, where the average diameter is taken.

CUBICAL CONTENTS OF ROUND TIMBER

L. ft.	Dia. 6	Dia. 7	Dia 8	Dia. 9	Dia. 10	Dia. 11	Dia. 12	Dia. 13
8	1.57	2.14	2.79	3.53	4	5	6	7
9	1.76	2.40	3.14	3.97	5	6	7	8
10	1.96	2.67	3.49	4.42	5	7	8	9
11	2.16	2.94	3.84	4.86	6	7	8	10
12	2.35	3.20	4.19	5.30	6	8	9	11
13	2.55	3.47	4.54	5.74	7	9	10	12
14	2.75	3.74	4.89	6.19	7	9	11	13
15	2.94	4.05	5.24	6.63	8	10	12	14
16	3.14	4.27	5.58	7.07	9	11	12	14
17	3.33	4.54	5.93	7.51	9	11	13	16
18	3.53	4.81	6.28	7.95	10	12	14	16
19	3.73	5.07	6.63	8.39	10	13	15	17
20	3.92	5.34	6.98	8.84	11	13	16	18
21	4.12	5.61	7.33	9.28	11	14	16	19
22	4.32	5.88	7.67	9.72	12	15	17	20
23	4.51	6.14	8.03	10.16	12	16	18	21
24	4.70	6.41	8.37	10.60	13	16	19	22
25	4.90	6.68	8.72	11.05	14	17	20	23
26	5.10	6.94	9.07	11.49	14	17	20	24
27	5.29	7.21	9.42	11.93	15	18	21	25
28	5.49	7.48	9.77	12.37	15	18	22	26
29	5.68	7.74	10.12	12.81	16	19	23	27
30	5.88	8.01	10.47	13.26	16	20	24	28
31	6.08	8.28	10.82	13.70	17	20	24	29
32	6.27	8.54	11.17	14.13	17	21	25	29
33	6.48	8.82	11.52	14.58	18	22	26	30
34	6.67	9.08	11.86	15.02	19	22	27	31
35	6.87	9.35	12.21	15.47	19	23	28	32
36	7.05	9.62	12.56	15.90	20	24	28	33

CUBICAL CONTENTS OF ROUND TIMBER

L ft.	Dia. 14	Dia. 15	Dia. 16	Dia. 17	Dia. 18	Dia. 19	Dia. 20	Dia. 21	Dia. 22
8	8	10	11	12	14	16	17	19	21
9	9	11	12	14	16	18	20	22	24
10	10	12	14	16	18	20	22	24	26
11	12	13	15	17	19	22	24	26	29
12	13	15	17	19	21	24	26	29	32
13	14	16	18	20	23	26	28	31	34
14	15	17	19	22	25	28	31	34	37
15	16	18	21	23	26	30	33	36	40
16	17	20	22	25	28	32	35	38	42
17	18	21	24	27	30	33	37	41	45
18	19	22	25	28	32	35	39	43	48
19	21	23	27	30	33	37	41	45	50
20	21	25	28	31	35	39	44	48	53
21	22	26	29	33	37	41	46	50	55
22	23	27	31	35	39	43	48	53	58
23	24	28	32	36	41	45	50	55	61
24	26	30	34	38	42	47	52	58	63
25	27	31	35	39	44	49	54	60	66
26	28	32	36	41	46	51	57	63	69
27	29	33	38	42	48	53	59	65	71
28	30	35	39	44	49	55	61	67	74
29	31	36	41	45	51	57	63	70	77
30	32	37	42	47	53	59	65	72	79
31	33	38	43	48	55	61	68	75	82
32	34	40	45	50	57	63	70	77	85
33	35	41	46	52	58	65	72	79	87
34	36	42	48	53	60	67	74	82	90
35	37	43	49	55	62	69	76	84	93
36	39	44	50	57	64	71	79	86	95

CUBICAL CONTENTS OF ROUND TIMBER

L. ft.	Dia. 23	Dia. 24	Dia. 25	Dia. 26	Dia. 27	Dia. 28	Dia. 29	Dia. 30
8	23	25	27	29	32	34	37	39
9	26	28	31	33	36	38	41	44
10	29	31	34	37	40	43	46	49
11	32	35	37	41	43	47	50	53
12	34	38	41	44	47	51	55	58
13	37	41	44	48	51	56	60	63
14	40	44	48	52	55	60	64	68
15	43	47	51	55	59	64	69	73
16	46	50	55	59	63	68	73	78
17	49	53	58	63	68	73	78	83
18	52	57	61	66	72	77	82	88
19	55	60	65	70	75	81	87	93
20	58	63	68	74	79	85	91	98
21	61	66	71	77	83	90	96	103
22	64	69	75	81	87	94	101	109
23	66	72	78	85	91	98	105	113
24	69	75	82	88	95	102	111	118
25	72	79	85	92	99	107	116	123
26	75	82	89	96	103	111	120	128
27	78	85	92	99	107	115	125	133
28	81	88	95	103	111	120	129	136
29	84	91	99	107	115	124	134	143
30	86	94	102	110	119	128	138	148
31	89	98	106	114	123	132	143	152
32	92	100	109	118	127	137	148	157
33	95	104	112	121	130	141	152	162
34	98	107	116	125	135	145	157	167
35	101	110	119	129	139	149	161	172
36	104	113	123	133	143	154	166	177

CUBICAL CONTENTS OF ROUND TIMBER

L. ft.	Dia. 31	Dia. 32	Dia. 33	Dia. 34	Dia. 35	Dia. 36	Dia. 37	Dia. 38
8	42	45	48	50	53	57	60	62
9	47	50	53	57	60	64	67	70
10	52	56	59	63	67	71	75	79
11	57	61	65	69	73	77	82	86
12	62	67	71	76	80	85	90	94
13	68	72	77	82	87	92	97	102
14	73	78	83	88	94	99	105	110
15	78	84	89	95	100	106	112	118
16	83	89	95	101	107	113	119	126
17	89	95	101	107	114	121	127	135
18	94	100	106	114	120	128	134	142
19	99	106	112	120	127	135	142	151
20	105	112	118	126	134	142	149	159
21	111	117	124	132	140	149	157	166
22	116	123	130	139	147	156	164	174
23	121	128	136	145	154	163	172	183
24	127	134	143	151	160	170	179	191
25	131	139	149	158	167	178	187	198
26	137	145	154	164	174	185	194	206
27	142	151	160	170	180	192	202	214
28	147	156	166	177	187	198	209	222
29	153	162	172	183	194	206	217	228
30	158	168	177	189	200	213	224	236
31	163	173	182	195	207	220	232	244
32	169	178	188	202	214	227	239	253
33	174	184	194	208	220	234	247	261
34	179	190	200	214	227	241	254	268
35	182	196	205	220	234	248	261	276
36	190	201	212	227	240	255	269	284

SHOWING THE CONTENTS OF STANDARD SAW-LOGS, FROM 10 IN. DIAM. TO 42

Dia.	Decimals	Inches	Dia.	Decimals	Inches
10	100	.277	27	729	2.020
11	121	.335	28	784	2.171
12	144	.399	29	841	2.330
13	169	.478	30	900	2.493
14	196	.543	31	961	2.662
15	225	.623	32	1024	2.836
16	256	.709	33	1089	3.016
17	289	.800	34	1156	3.202
18	324	.897	35	1225	3.400
19	361	1.000	36	1296	3.590
20	400	1.108	37	1369	3.792
21	441	1.221	38	1444	4.000
22	484	1.341	39	1521	4.213
23	529	1.465	40	1600	4.432
24	576	1.595	41	1681	4.656
25	625	1.731	42	1764	4.886
26	676	1.872			

REMARKS.—In most lumbering districts, where *piece* lumber is manufactured the *standard* measure for logs is 19 inches diameter, and 13 feet long, which, it will be seen, gives 361 decimals, or 100 standard inches. Thus, $19 \times 19 = 361$, and $361 \div 361 = 1.00$, which is the standard. If the log exceeds this standard, either in length or diameter the surplus is reckoned as the decimal parts of another log.

EXAMPLE —What are the standard contents of a log 23 inches diameter and 13 feet long? $23 \div 23 = 529$, and $529 \div 361 = 1.46$, which is one log and 46-100 of another.

NOTE.—Multiply the standard inches given for a log of any given diameter by the number of logs of the same diameter the product will be the measure for such number of logs ⁀ The diameter is to be the average measure. taken at the smallest end inside the bark.

LOG TABLE

ROUND LOGS REDUCED TO INCH BOARD MEASURE
BY DOYLE'S RULE

The length of the log, in feet, will be found in
the left hand column of the table, and the diameter at the top of the page.   To find the number of feet of *square edged* board which a log will
produce when sawed, take the length, in feet,
in the left hand column of the table, and its diameter in inches at the top of the page; trace the
two columns of figures until they meet, and you
will have the amount.

Thus, a log
which is 18 ft.
long, and 16 in.
in d i a m e t e r,
gives, at the
right o f t h e
length, and di-
rectly u n d e r
the diameter,
162 ft., while
one 36 feet in
length and 18

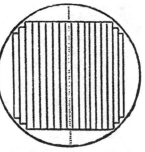

inches diameter gives 440 feet, fractions omitted.

The diagram shows the manner of sawing up
the logs into boards, and the table indicates the
number of feet in any given log.

If logs are more than 50 feet long, add the
measurements of shorter lengths *of same diameter*, to make the length desired, viz.: if 65 feet

long by 30 inches diameter is wanted, add 30 ft. long by 30 inches in diameter, to 35 ft. long by same diameter—1266 + 1479 = 2745 feet.

The measurements of logs of larger *diameters* than those given in the tables cannot be obtained in this way.

It is customary in measuring logs to take the diameter in the middle of the log, inside the bark. This is obtained by taking the diameter at each end of the log, adding them together and dividing by 2. It is usual to allow, on account of the bark, for oak 1-10th or 1-12th part of the circumference; for beech, ash, etc., less should be allowed.

Logs are seldom exactly round or perfectly straight, besides having many irregularities covered by the bark, hence allowance should be made to the purchaser.

Logs that are less than 10 inches in diameter have very little left after taking off the slab and saw kerf; unless valuable timber they would be worth more for *wood;* this should be considered by farmers bringing small logs to market, for they often get less for them as *logs* than they would if sold for *fuel.*

REMARKS.—In this revised edition of Scribner's Book we continue to use Doyle's Log Rule. From repeated letters and opinions of old saw-mill men and large lumber dealers throughout the country, universally using and approving our book, we are satisfied that "Doyle's Log Rule" gives fair and honest measurements, alike just to both buyer and seller. We are aware that there are several log tables in the market, no two being alike but each claiming to be the only correct one. As Scribner's book has had a much larger sale than all combined books of its kind ever published, we are willing to leave the public to decide on the merits of the log tables.

ROUND LOGS REDUCED TO INCH BOARD MEASURE—DOYLE'S RULE

L. ft.	Dia. 8	Dia. 9	Dia. 10	Dia. 11	Dia. 12	Dia. 13	Dia. 14	Dia. 15	Dia. 16
8	8	12	18	24	32	40	50	60	72
9	9	14	20	28	36	46	56	68	81
10	10	16	23	31	40	50	62	75	90
11	11	17	25	34	44	55	69	83	99
12	12	19	27	37	48	61	75	91	108
13	13	20	29	40	52	66	81	98	117
14	14	22	32	43	56	71	88	106	126
15	15	23	34	46	60	76	94	113	135
16	16	25	36	49	64	81	100	121	144
17	17	27	38	52	68	86	106	128	153
18	18	28	41	55	72	91	112	136	162
19	19	30	43	58	76	96	119	143	171
20	20	31	46	61	80	101	125	151	180
21	21	33	48	64	84	106	131	158	189
22	22	34	50	67	88	111	137	166	198
23	23	36	52	70	92	116	144	174	207
24	24	37	54	74	96	122	150	181	216
25	25	39	56	77	100	127	156	189	225
26	26	41	59	80	104	132	163	196	234
27	27	42	61	83	108	137	169	204	243
28	28	44	63	86	112	142	175	212	252
29	29	45	65	89	116	147	182	219	261
30	30	47	68	92	120	152	188	226	270
31	31	48	70	95	124	157	193	234	279
32	32	50	72	98	128	162	200	242	288
33	33	52	74	101	132	167	206	249	297
34	34	53	77	104	136	172	212	256	306
35	35	55	79	107	140	177	219	265	315
36	36	56	81	110	144	182	225	272	324
37	37	58	83	113	148	187	231	280	333
38	38	59	85	116	152	192	237	287	342
39	39	61	88	119	156	197	243	295	351
40	40	62	90	122	160	202	250	302	360

ROUND LOGS REDUCED TO INCH BOARD MEASURE—DOYLE'S RULE

L. ft.	Dia. 17	Dia. 18	Dia. 19	Dia. 20	Dia. 21	Dia. 22	Dia. 23	Dia. 24
8	84	98	112	128	144	162	180	200
9	95	110	127	144	163	182	203	225
10	106	122	141	160	181	202	226	250
11	116	135	155	176	199	223	248	275
12	127	147	169	192	217	243	271	300
13	137	159	183	208	235	263	293	325
14	148	171	197	224	253	283	313	350
15	158	184	211	240	271	303	336	375
16	169	196	225	256	289	324	359	400
17	180	208	239	272	307	344	383	425
18	190	220	253	288	325	364	406	450
19	201	233	267	304	343	384	429	475
20	211	245	280	320	361	404	452	500
21	222	257	295	336	379	425	473	525
22	232	269	309	352	397	445	496	550
23	243	282	323	368	415	465	519	575
24	253	294	338	384	433	486	541	600
25	264	306	351	400	451	506	562	625
26	275	318	366	416	470	526	586	650
27	285	331	380	432	488	546	606	675
28	296	343	394	448	506	566	626	700
29	306	355	408	464	524	586	649	725
30	317	367	421	480	542	606	672	750
31	327	380	436	496	560	627	695	775
32	338	392	450	512	578	648	718	800
33	349	404	464	528	596	668	742	825
34	359	416	478	544	614	688	766	850
35	370	429	492	560	632	708	789	875
36	380	441	506	576	650	729	812	900
37	391	453	520	592	668	749	835	925
38	401	465	534	608	686	769	857	950
39	412	478	548	624	704	790	880	975
40	422	490	562	640	722	810	903	1000

ROUND LOGS REDUCED TO INCH
BOARD MEASURE—DOYLE'S RULE

L ft.	Dia. 25	Dia. 26	Dia. 27	Dia. 28	Dia. 29	Dia. 30	Dia. 31	Dia. 32
8	220	242	264	288	312	338	364	392
9	248	272	297	324	352	380	410	441
10	276	302	330	360	391	422	456	490
11	303	334	363	396	430	465	502	539
12	331	363	397	432	469	507	547	588
13	358	393	430	468	508	549	592	637
14	386	423	463	504	547	591	638	686
15	413	458	496	540	586	633	683	735
16	441	484	530	576	625	676	729	784
17	469	514	563	612	664	718	774	833
18	496	544	596	648	703	761	820	882
19	524	575	630	684	742	803	865	931
20	551	605	661	720	782	845	912	980
21	579	635	693	756	820	887	957	1029
22	606	665	726	792	860	930	1004	1078
23	634	696	760	828	898	972	1049	1127
24	661	726	794	864	938	1014	1094	1176
25	689	756	827	900	977	1056	1139	1225
26	717	786	860	936	1016	1098	1184	1274
27	744	817	893	972	1055	1140	1230	1323
28	772	847	926	1008	1094	1182	1276	1372
29	799	877	959	1044	1133	1224	1321	1421
30	827	907	992	1080	1172	1266	1366	1470
31	854	938	1026	1116	1211	1309	1412	1519
32	882	968	1060	1152	1250	1352	1458	1568
33	910	998	1093	1188	1289	1394	1503	1617
34	937	1028	1126	1224	1328	1436	1548	1666
35	965	1059	1159	1260	1367	1479	1594	1715
36	992	1089	1192	1296	1406	1522	1640	1764
37	1020	1119	1223	1332	1445	1563	1686	1813
38	1047	1149	1256	1368	1484	1606	1731	1862
39	1075	1180	1289	1404	1523	1648	1778	1911
40	1102	1210	1322	1440	1562	1690	1822	1960

ROUND LOGS REDUCED TO INCH
BOARD MEASURE—DOYLE'S RULE

L. Ft.	Dia. 33	Dia. 34	Dia. 35	Dia. 36	Dia. 37	Dia. 38	Dia. 39
8	420	450	480	512	544	578	612
9	473	506	540	576	613	650	689
10	526	562	601	640	681	723	765
11	578	619	661	704	749	795	842
12	631	675	721	768	817	867	910
13	683	731	781	832	884	939	996
14	736	787	841	896	953	1011	1070
15	789	844	901	960	1021	1083	1149
16	841	900	961	1024	1089	1156	1225
17	894	956	1021	1088	1157	1228	1302
18	946	1012	1081	1152	1225	1300	1379
19	999	1069	1141	1216	1293	1372	1455
20	1051	1125	1202	1280	1361	1446	1530
21	1104	1181	1261	1344	1430	1518	1607
22	1156	1237	1322	1408	1497	1590	1684
23	1209	1293	1381	1472	1566	1662	1761
24	1262	1350	1442	1536	1634	1734	1838
25	1314	1406	1501	1600	1702	1806	1915
26	1367	1462	1562	1664	1768	1878	1992
27	1420	1518	1622	1728	1838	1950	2067
28	1472	1575	1682	1792	1906	2022	2144
29	1524	1631	1742	1856	1974	2095	2221
30	1577	1687	1802	1920	2042	2166	2298
31	1629	1743	1862	1984	2110	2239	2373
32	1682	1800	1922	2048	2178	2312	2450
33	1735	1856	1982	2112	2246	2386	2526
34	1787	1912	2042	2176	2314	2456	2604
35	1840	1968	2102	2240	2383	2529	2681
36	1892	2025	2162	2304	2450	2601	2756
37	1945	2081	2222	2368	2518	2673	2833
38	1998	2138	2282	2432	2586	2745	2908
39	2050	2194	2342	2496	2654	2818	2986
40	2102	2250	2402	2560	2722	2890	3062

ROUND LOGS REDUCED TO INCH
BOARD MEASURE—DOYLE'S RULE

L. Ft.	Dia. 40	Dia. 41	Dia. 42	Dia. 43	Dia. 44	Dia. 45	Dia. 46
8	648	684	722	761	800	840	882
9	729	770	812	856	900	946	992
10	810	856	902	951	1000	1051	1103
11	891	941	993	1046	1100	1156	1213
12	972	1027	1083	1141	1200	1261	1323
13	1053	1112	1173	1237	1300	1366	1434
14	1134	1198	1264	1331	1400	1471	1544
15	1215	1284	1354	1426	1500	1576	1654
16	1296	1369	1444	1521	1600	1681	1764
17	1377	1455	1534	1616	1700	1786	1874
18	1458	1540	1625	1711	1800	1891	1985
19	1539	1626	1715	1806	1900	1996	2096
20	1620	1711	1805	1902	2000	2102	2206
21	1701	1797	1895	1997	2100	2207	2316
22	1782	1882	1986	2091	2200	2312	2426
23	1863	1968	2076	2187	2300	2416	2536
24	1944	2053	2166	2282	2400	2522	2646
25	2025	2139	2256	2376	2500	2627	2757
26	2106	2225	2346	2472	2600	2732	2868
27	2187	2310	2437	2567	2700	2837	2978
28	2268	2396	2527	2662	2800	2942	3088
29	2349	2481	2617	2756	2900	3047	3198
30	2430	2567	2707	2852	3000	3152	3308
31	2511	2652	2798	2946	3100	3257	3418
32	2592	2738	2888	3042	3200	3362	3528
33	2673	2824	2978	3137	3300	3467	3638
34	2754	2909	3068	3232	3400	3572	3748
35	2835	2995	3159	3327	3500	3677	3858
36	2916	3080	3249	3423	3600	3782	3969
37	2997	3166	3339	3517	3700	3887	4079
38	3078	3251	3429	3612	3800	3992	4190
39	3159	3337	3520	3707	3900	4097	4300
40	3240	3423	3610	3802	4000	4202	4410

ROUND LOGS REDUCED TO INCH BOARD MEASURE—DOYLE'S RULE

L. Ft.	Dia. 47	Dia. 48	Dia. 49	Dia. 50	Dia. 51	Dia. 52	Dia. 53
8	925	968	1013	1058	1105	1152	1200
9	1040	1089	1139	1190	1243	1296	1350
10	1155	1210	1266	1322	1380	1440	1500
11	1271	1331	1392	1455	1519	1584	1650
12	1387	1452	1519	1587	1657	1728	1801
13	1502	1573	1645	1719	1795	1872	1951
14	1618	1694	1772	1850	1933	2016	2101
15	1734	1815	1898	1984	2071	2160	2251
16	1849	1936	2025	2116	2209	2304	2401
17	1964	2057	2152	2248	2347	2448	2551
18	2080	2178	2278	2380	2485	2592	2701
19	2195	2299	2403	2513	2623	2736	2851
20	2312	2420	2530	2645	2761	2880	3001
21	2427	2541	2657	2777	2899	3024	3151
22	2542	2662	2784	2909	3037	3168	3301
23	2658	2783	2911	3041	3175	3312	3451
24	2774	2904	3038	3174	3313	3456	3601
25	2889	3025	3164	3306	3451	3600	3752
26	3004	3146	3290	3438	3590	3744	3902
27	3120	3267	3417	3571	3728	3888	4052
28	3236	3388	3544	3701	3866	4032	4202
29	3351	3509	3670	3835	4004	4176	4352
30	3467	3630	3796	3968	4142	4320	4502
31	3583	3751	3923	4100	4280	4464	4652
32	3698	3872	4050	4232	4418	4608	4802
33	3812	3993	4177	4364	4556	4752	4952
34	3928	4114	4303	4497	4694	4896	5102
35	4045	4235	4429	4629	4832	5040	5252
36	4161	4356	4556	4761	4970	5184	5402
37	4276	4477	4683	4893	5108	5320	5552
38	4391	4598	4809	5025	5246	5472	5702
39	4507	4719	4936	5158	5384	5616	5852
40	4622	4840	5062	5290	5522	5716	6002

ROUND LOGS REDUCED TO INCH
BOARD MEASURE—DOYLE'S RULE

L. Ft.	Dia. 54	Dia. 55	Dia. 56	Dia. 57	Dia. 58	Dia. 59	Dia. 60
8	1250	1300	1352	1404	1458	1512	1568
9	1406	1463	1521	1580	1640	1702	1764
10	1562	1626	1690	1756	1822	1891	1960
11	1719	1788	1859	1931	2005	2080	2156
12	1875	1951	2028	2107	2187	2269	2352
13	2031	2113	2197	2282	2369	2458	2548
14	2187	2276	2366	2458	2551	2647	2744
15	2344	2438	2535	2633	2734	2836	2940
16	2500	2601	2704	2809	2916	3025	3136
17	2656	2763	2873	2985	3098	3214	3332
18	2812	2926	3042	3160	3280	3403	3528
19	2969	3088	3211	3336	3463	3592	3724
20	3125	3251	3380	3511	3645	3781	3920
21	3281	3413	3549	3687	3827	3970	4116
22	3437	3576	3718	3862	4009	4159	4312
23	3594	3738	3887	4038	4192	4348	4508
24	3750	3901	4056	4213	4374	4537	4704
25	3906	4063	4225	4389	4556	4727	4900
26	4062	4226	4394	4565	4738	4916	5096
27	4219	4388	4563	4740	4921	5105	5292
28	4375	4552	4732	4916	5103	5294	5488
29	4531	4714	4901	5091	5285	5483	5684
30	4687	4877	5070	5267	5467	5672	5880
31	4844	5039	5239	5442	5650	5861	6076
32	5000	5202	5408	5618	5832	6050	6272
33	5156	5364	5577	5794	6014	6239	6468
34	5314	5527	5746	5969	6196	6428	6664
35	5469	5689	5915	6145	6379	6617	6860
36	5625	5852	6084	6320	6561	6806	7056
37	5781	6014	6253	6496	6743	6995	7252
38	5937	6177	6422	6671	6925	7184	7448
39	6094	6339	6591	6847	7108	7373	7644
40	6250	6502	6760	7022	7290	7562	7840

NUMBER OF FEET IN LENGTH OF THE FOLLOWING DIMENSIONS OF TIMBER REQUIRED TO MAKE 1,000 FEET OF BOARD MEASURE

Size	No. of feet in length to make 1,000 feet B.M.	Size	No. of feet in length to make 1,000 feet B.M.	Size	No. of feet in length to make 1,000 feet B.M.
2 x 6	1000.	5 x 6	400.	10 x 10	120.
2 x 7	857.2	5 x 7	342.10	10 x 11	109.1
2 x 8	750.	5 x 8	300.	10 x 12	100.
2 x 9	666.8	5 x 9	266.8	11 x 11	99.2
2 x 10	600.	5 x 10	240.	11 x 12	90.9
2 x 11	545.6	5 x 11	218.2	12 x 12	83.4
2 x 12	500.	5 x 12	200.	12 x 14	71.5
2½ x 5	960.	6 x 6	333.4	12 x 16	62.5
2½ x 6	800.	6 x 7	285.8	12 x 18	55.6
2½ x 7	685.9	6 x 8	250.	12 x 20	50.
2½ x 8	600.	6 x 9	222.2	13 x 14	66.11
2½ x 9	533.4	6 x 10	200.	14 x 16	53.7
2½ x 10	480.	6 x 11	181.10	15 x 18	44.5
3 x 5	800.	6 x 12	166.8	16 x 18	41.8
3 x 6	666.8	7 x 7	244.11	16 x 20	37.6
3 x 7	571.5	7 x 8	214.3	18 x 20	33.4
3 x 8	500.	7 x 9	190.6	18 x 24	27.10
3 x 9	444.4	7 x 10	171.5	20 x 20	30.
3 x 10	400.	7 x 11	155.10	20 x 24	25.
3 x 11	363.7	7 x 12	142.10	22 x 24	22.8
3 x 12	333.4	8 x 8	187.6	30 x 40	10.
4 x 5	600.	8 x 9	166.8	36 x 36	9.3
4 x 6	500.	8 x 10	150.		
4 x 7	428.7	8 x 11	136.4	*Explanation.—*	
4 x 8	375.	8 x 12	125.	If 2 x 6 it takes	
4 x 9	333.4	9 x 9	148.2	1000 feet long;	
4 x 10	300.	9 x 10	133.4	while 8 x 10 it	
4 x 11	272.8	9 x 11	121.3	takes 150 feet	
4 x 12	250.	9 x 12	111.2	long	

PRICE PER FT. OF STANDARD LOGS
OF 300 FEET

Fractions omitted, or if less than ½, nothing; if over, 1 cent

No. of Feet	Per Log $1.00	Per Log $1.25	Per Log $1.50	Per Log $1.75	Per Log $2.00	Per Log $2.25	Per Log $2.50	Per Log $2.75	Per Log $3.00	Per Log $3.25
5	.02	.02	.02	.03	.03	.04	.04	.04	.05	.05
6	.02	.02	.03	.03	.04	.04	.05	.05	.06	.06
7	.02	.02	.03	.04	.04	.05	.06	.06	.07	.07
8	.03	.03	.04	.04	.05	.06	.07	.07	.08	.08
9	.03	.04	.04	.05	.06	.06	.07	.08	.09	.10
10	.03	.04	.05	.06	.07	.07	.08	.09	.10	.11
15	.05	.06	.07	.09	.10	.11	.12	.13	.15	.16
20	.07	.08	.10	.11	.13	.15	.16	.18	.20	.22
25	.08	.10	.12	.14	.17	.19	.21	.23	.25	.27
30	.10	.12	.15	.17	.20	.22	.25	.27	.30	.32
35	.12	.14	.16	.20	.23	.26	.29	.32	.35	.38
40	.13	.17	.20	.23	.26	.30	.33	.37	.40	.43
45	.15	.19	.22	.25	.30	.33	.37	.41	.44	.49
50	.17	.21	.25	.29	.33	.37	.40	.46	.50	.54
55	.18	.23	.27	.32	.37	.41	.45	.50	.55	.59
60	.20	.25	.30	.35	.40	.45	.50	.55	.60	.65
65	.22	.27	.32	.38	.43	.48	.53	.59	.65	.70
70	.23	.29	.35	.41	.47	.53	.58	.64	.70	.76
75	.25	.31	.37	.44	.50	.56	.60	.68	.75	.81
80	.27	.33	.40	.47	.53	.59	.67	.73	.79	.86
85	.28	.35	.42	.49	.56	.63	.71	.77	.85	.92
90	.30	.37	.45	.52	.00	.67	.75	.82	.90	.97
95	.32	.39	.47	.54	.63	.71	.79	.87	.95	1.02
100	.33	.42	.50	.58	.67	.75	.83	.92	1.00	1.08

IN SOME sections of the country logs are bought and sold by the log, the log to contain what is called *standard measurement*, i.e., it must be 12 ft. long and 24 in. diameter, measured at the small end inside the bark, and contain 300 feet, board measure.

PRICE PER FT. OF STANDARD LOGS OF 300 FEET

No. of Feet	Per Log $3.50	Per Log $3.75	Per Log $4.00	Per Log $4.50	Per Log $5.00	Per Log $5.50	Per Log $6.00
5	.06	.07	.07	.07	.08	.09	.10
6	.07	.07	.08	.09	.10	.10	.12
7	.08	.09	.10	.10	.12	.12	.14
8	.09	.10	.11	.12	.13	.15	.17
9	.10	.11	.12	.13	.15	.16	.18
10	.12	.12	.13	.15	.17	.18	.20
15	.17	.18	.20	.22	.25	.27	.30
20	.23	.25	.27	.30	.33	.37	.40
25	.29	.31	.33	.37	.42	.46	.50
30	.35	.37	.40	.45	.50	.55	.60
35	.41	.44	.47	.52	.58	.64	.70
40	.47	.50	.53	.60	.67	.73	.80
45	.52	.56	.60	.67	.75	.82	.90
50	.58	.62	.67	.75	.83	.92	1.00
55	.64	.68	.73	.82	.92	1.01	1.10
60	.70	.75	.80	.90	1.00	1.10	1.20
65	.76	.81	.87	.97	1.08	1.19	1.30
70	.82	.88	.93	1.05	1.17	1.28	1.40
75	.87	.93	1.00	1.12	1.25	1.37	1.50
80	.93	1.00	1.07	1.20	1.33	1.47	1.60
85	.99	1.06	1.13	1.27	1.42	1.56	1.70
90	1.05	1.12	1.20	1.35	1.50	1.65	1.80
95	1.11	1.19	1.27	1.42	1.58	1.74	1.90
100	1.17	1.25	1.33	1.50	1.67	1.83	2.00

The price will be found at top of page, the number of ft. in left hand column; trace across the page until you come under the price per log and you will have the required amount.

EXAMPLE—To determine price of the odd feet, suppose a person has sold a number of logs which measured altogether 1,675 feet, there would be 5 logs of 300 feet and 175 feet over, how much would they come to at $3.50 per log?   5 logs of 300 feet at $3.50 would be $17.50

100 feet by the table would come to $1.17

75   .87

Making for 5 logs of 1,675 ft. at $3.50 per stand. log of 300 ft., frac'ns omitted

2.04

$19.54

THE ABOVE table is designed to aid farmers and small dealers who are in the habit of buying or selling logs by standard measurement of 300 feet to the log, to determine what the odd feet come to at so much per log.

LOG TALLY

CALCULATOR

BY THIS TABLE

One or more tallies may be quickly calculated.

For example, you have 12 tallies of 18 ft each, you wish to know the number of ft. they contain; find the number 12 in the left hand column and the 18 in top column, trace each to their meeting, when it will be seen that there are 216 feet in the tally.

	1	2	3	4	5
	2	4	6	8	10
	3	6	9	12	15
	4	8	12	16	20
	5	10	15	20	25
	6	12	18	24	30
	7	14	21	28	35
	8	16	24	32	40
	9	18	27	36	45
	10	20	30	40	50
	11	22	33	44	55
	12	24	36	48	60
	13	26	39	52	65
	14	28	42	56	70
	15	30	45	60	75
	16	32	48	64	80
	17	34	51	68	85
	18	36	54	72	90
	19	38	57	76	95
	20	40	60	80	100
	21	42	63	84	105
	22	44	66	88	110
	23	46	69	92	115
	24	48	72	96	120
	25	50	75	100	125

LOG TALLY

CALCULATOR

	6	7	8	9	10	11	12	13	14	15
2	12	14	16	18	20	22	24	26	28	30
3	18	21	24	27	30	33	36	39	42	45
4	24	28	32	36	40	44	48	52	56	60
5	30	35	40	45	50	55	60	65	70	75
6	36	42	48	54	60	66	72	78	84	90
7	42	49	56	63	70	77	84	91	98	105
8	48	56	64	72	80	88	96	104	112	120
9	54	63	72	81	90	99	108	117	126	135
10	60	70	80	90	100	110	120	130	140	150
11	66	77	88	99	110	121	132	143	154	165
12	72	84	96	108	120	132	144	156	168	180
13	78	91	104	117	130	143	156	169	182	195
14	84	98	112	126	140	154	168	182	196	210
15	90	105	120	135	150	165	180	195	210	225
16	96	112	128	144	160	176	192	208	224	240
17	102	119	136	153	170	187	204	221	238	255
18	108	126	144	162	180	198	216	234	252	270
19	114	133	152	171	190	209	228	247	266	285
20	120	140	160	180	200	220	240	260	280	300
21	126	147	168	189	210	231	252	273	294	315
22	132	154	176	198	220	242	264	286	308	330
23	138	161	184	207	230	253	276	299	322	345
24	144	168	192	216	240	264	288	312	336	360
25	150	175	200	225	250	275	300	325	350	375

LOG TALLY

CALCULATOR

	16	17	18	19	20	21	22	23	24	25
2	32	34	36	38	40	42	44	46	48	50
3	48	51	54	57	60	63	66	69	72	75
4	64	68	72	76	80	84	88	92	96	100
5	80	85	90	95	100	105	110	115	120	125
6	96	102	108	114	120	126	132	138	144	150
7	112	119	126	133	140	147	154	161	168	175
8	128	136	144	152	160	168	176	184	192	200
9	144	153	162	171	180	189	198	207	216	225
10	160	170	180	190	200	210	220	230	240	250
11	176	187	198	209	220	231	242	253	264	275
12	192	204	216	228	240	252	264	276	288	300
13	208	221	234	247	260	273	286	299	312	325
14	224	238	252	266	280	294	308	322	336	350
15	240	255	270	285	300	315	330	345	360	375
16	256	272	288	304	320	336	352	368	384	400
17	272	289	306	323	340	357	374	391	408	425
18	288	306	324	342	360	378	396	414	432	450
19	304	323	342	361	380	399	418	437	456	475
20	320	340	360	380	400	420	440	460	480	500
21	336	357	378	399	420	441	462	483	504	525
22	352	374	396	418	440	462	484	506	528	550
23	368	391	414	437	460	483	506	529	552	575
24	384	408	432	456	480	504	528	552	576	600
25	400	425	450	475	500	525	550	575	600	625

TABLE OF SPECIFIC GRAVITY AND WEIGHT OF DIFFERENT WOODS

CAPACITY OF CISTERNS

TIMBER	Specific Gravity	Lbs. per C. Foot	Bottom ft.	Bottom in.	St've ft.	St've in.	Cap. Bbls.
			3	6	3	6	7
			4		4		11
			4		4	8	13
Oak, dry........	.625	39.06	4		5	4	15
Oak, green......	1.113	69.56	4	6	4		14
Beech, dry......	.69	43.12	4		4	8	16
Maple..........	.795	49.68	4	6	5	4	18
Sycamore, dry...	.590	36.87	5		4		18
" green.	.645	40.31	5		4	8	20
Chestnut, dry...	.535	33.45	5		5	4	22
" green..	.875	54.68	5		6		26
Ash, dry........	.845	52.81	5	6	4		22
Elm, dry........	.588	36.75	5	6	4	8	25
" green..	.940	58.75	5	6	5	4	27
Walnut, green...	.920	57.50	5	6	6		31
" dry.....	.616	38.50	6		4	8	30
Poplar..........	.421	26.31	6		5	4	32
Cedar..........	.560	35.	6		6		37
" dry....	.453	28.31	6		7		46
Lignum Vitæ....	1.333	83.31	6	6	5	4	38
Pine	.368	23.	6		6	6	43
" pitch......	.936	58.5	6		6	7	51
Mahogany, dry..:	.852	53.30	6		6	8	61
Willow, green....	.619	38.68	7		5	4	44
" dry......	.486	30.37	7		6		50
Water..........	1.	62.50	7		7		59

ACCURATE WOOD MEASURER

LENGTH EIGHT FEET.

Width.		Height in ft.				Height in inches.										
ft.	in.	1	2	3	4	1	2	3	4	5	6	7	8	9	10	11
2	6	20	40	60	80	2	3	5	7	8	10	12	13	15	17	18
	7	21	41	62	82	2	3	5	7	8	10	12	14	15	17	18
	8	21	42	64	85	2	4	5	7	9	11	13	14	16	18	20
	9	22	44	66	88	2	4	6	8	9	11	13	15	17	18	20
	10	23	45	68	91	2	4	6	8	10	11	13	15	17	19	21
	11	23	47	70	94	2	4	6	8	10	12	14	15	17	19	21
3	0	24	48	72	96	2	4	6	8	10	12	14	16	18	20	22
	1	25	49	74	99	2	4	6	8	10	12	14	17	18	20	22
	2	25	51	76	101	2	4	6	8	10	13	15	17	19	21	23
	3	26	52	78	104	2	4	7	9	11	13	15	17	20	22	24
	4	27	53	80	107	2	5	7	9	11	14	16	18	20	23	25
	5	27	55	82	109	2	5	7	9	11	14	16	18	20	23	25
	6	28	56	84	112	2	5	7	9	12	14	16	19	21	23	26
	7	29	57	86	115	3	5	8	10	12	15	17	19	22	24	27
	8	29	59	88	117	3	5	8	10	12	15	17	19	22	24	27
	9	30	60	90	120	3	5	8	10	13	15	18	20	23	26	28
	10	31	61	92	123	3	5	8	10	13	16	18	21	23	26	29
	11	31	63	94	125	3	5	8	10	13	16	18	21	23	26	29
4	0	32	64	96	128	3	5	8	11	13	16	19	21	24	27	29

EXPLANATION.—Find the width of the load in the left hand column of the table; then move to the right, on the *same line*, till you come under the height in *feet*, and you have the contents in feet; then move to the right, on the same line, till you come to the height in *inches*, and you will have the *additional* contents in feet for the height in inches. The *sum* of these two gives the true contents in feet. For loads 12 feet long, add one-half, and for 4 feet, subtract one-half.

EXAMPLE.—If a load of wood be 2 feet 10 inches wide, and 3 feet 7 inches high, what are the contents? Against 2 feet 10 inches, and under 3 feet, stands 68; and under 7 inches, at the top, stands 13; then 13+68=81, the true contents in feet.

PRICE OF WOOD PER CORD

Ft.	$1.50	$1.75	$2.00	$2.25	$2.50	$2.75
1	.01	.01	.01	.02	.02	.02
2	.02	.02	.03	.03	.04	.04
3	.03	.04	.04	.05	.06	.06
4	.05	.06	.06	.07	.08	.09
5	.06	.07	.08	.09	.10	.11
6	.07	.08	.09	.11	.12	.13
7	.08	.10	.11	.12	.14	.15
8	.09	.11	.12	.14	.16	.18
16	.19	.22	.25	.28	.31	.35
24	.28	.33	.37	.42	.47	.52
32	.38	.44	.50	.56	.63	.69
40	.47	.55	.63	.70	.78	.86
48	.56	.66	.75	.84	.94	1.03
56	.61	.77	.88	.98	1.09	1.20
64	.75	.88	1.00	1.13	1.25	1.38
72	.84	.98	1.13	1.27	1.41	1.55
80	.94	1.09	1.25	1.41	1.56	1.72
84	.98	1.15	1.31	1.48	1.64	1.81
88	1.03	1.20	1.38	1.55	1.72	1.89
92	1.08	1.26	1.44	1.62	1.80	1.98
96	1.13	1.31	1.50	1.69	1.88	2.06
104	1.22	1.42	1.63	1.83	2.03	2.23
112	1.31	1.53	1.75	1.97	2.19	2.41
120	1.41	1.64	1.88	2.11	2.34	2.58
128	1.50	1.75	2.00	2.25	2.50	2.75

NOTE.—If the price of wood is wanted at a *less price* than is shown in these tables, take one-half of twice the price—i.e., if at 75 cents per cord, take one-half of what is shown for $1.50 per cord, if at $1.00 take one-half of $2.00, etc.

PRICE OF WOOD PER CORD

Ft.	$3.00	$3.25	$3.50	$4.00	$4.50	$5.00
1	.02	.02	.02	.03	.03	.03
2	.05	.05	.05	.06	.07	.07
3	.07	.07	.08	.09	.10	.11
4	.09	.10	.10	.12	.14	.15
5	.12	.13	.13	.15	.17	.19
6	.14	.15	.16	.18	.21	.23
7	.16	.17	.19	.21	.24	.27
8	.19	.20	.21	.24	.28	.31
16	.37	.40	.43	.49	.56	.62
24	.56	.61	.65	.75	.84	.93
32	.75	.81	.87	1.00	1.12	1.25
40	.94	1.02	1.09	1.25	1.40	1.56
48	1.12	1.22	1.31	1.50	1.68	1.87
56	1.31	1.42	1.53	1.75	1.96	2.18
64	1.50	1.62	1.75	2.00	2.25	2.50
72	1.69	1.83	1.96	2.25	2.53	2.81
80	1.88	2.03	2.18	2.50	2.81	3.13
84	1.97	2.13	2.29	2.62	2.95	3.28
88	2.06	2.23	2.40	2.75	3.09	3.43
92	2.15	2.33	2.51	2.87	3.23	3.59
96	2.25	2.44	2.62	3.00	3.37	3.75
104	2.44	2.64	2.84	3.25	3.65	4.05
112	2.62	2.84	3.06	3.50	3.93	4.38
120	2.81	3.05	3.28	3.75	4.21	4.68
128	3.00	3.25	3.49	4.00	4.50	5.00

EXPLANATION.—Find the number of feet in the left hand column of the table; then the price in dollars and cents at the top of the page, and trace the line and column until they meet, and you will find the amount in dollars and cents.

PRICE OF WOOD PER CORD

Ft.	$5.50	$6.00	$6.50	$7.00	$7.50	$8.00
1	.04	.04	.05	.05	.05	.06
2	.08	.09	.10	.10	.11	.12
3	.12	.14	.15	.16	.17	.18
4	.17	.18	.20	.21	.23	.25
5	.21	.23	.25	.27	.29	.31
6	.25	.28	.30	.32	.35	.37
7	.30	.32	.35	.38	.41	.43
8	.34	.37	.40	.43	.46	.50
16	.68	.74	.81	.87	.93	1.00
24	1.03	1.12	1.22	1.31	1.41	1.50
32	1.37	1.50	1.62	1.75	1.87	2.00
40	1.72	1.87	2.03	2.19	2.34	2.50
48	2.06	2.25	2.44	2.62	2.81	3.00
56	2.40	2.62	2.84	3.06	3.28	3.50
64	2.75	3.00	3.25	3.50	3.75	4.00
72	3.09	3.37	3.65	3.93	4.28	4.50
80	3.43	3.74	4.06	4.37	4.68	5.00
84	3.60	3.94	4.26	4.59	4.92	5.25
88	3.78	4.12	4.47	4.81	5.16	5.50
92	3.95	4.30	4.67	5.03	5.40	5.75
96	4.12	4.49	4.87	5.25	5.62	6.00
104	4.47	4.87	5.28	5.69	6.09	6.50
112	4.80	5.24	5.69	6.12	6.56	7.00
120	5.15	5.62	6.09	6.56	7.03	7.50
128	5.50	6.00	6.50	7.00	7.50	8.00

EXAMPLE.—If a load of wood contains 96 feet, at two dollars and fifty cents per cord—first find the amount of 96 feet, which is $1.88; and then add the value of two feet, 4 cents, making $1.92. So of all similar examples.

PRICE OF LUMBER TABLE

THIS table will be found very convenient to persons dealing in lumber to ascertain, at a glance, how much any number of feet come to at a given price per thousand feet, board measure. The price will be found at the top of the page, the number of feet in the left hand column; trace from the number of feet across the page until you come under the price and you will have the sum sought. In making the table where fractions occur, if half and over, one is added; if less, nothing.

EXAMPLE.—Suppose you wish to know what 700 feet of lumber comes to at $3.00 per thousand feet. Look at the top of the page for the price, then trace down the left hand column for the 700 feet, then trace across the page until under the price, and you have $2.10, being the price of 700 feet at $3.00 per thousand—while 125 feet at $20.00 per thousand comes to $2.50. If 715 feet at $35.00 is wanted, the table shows that 700 feet comes to................$24.50
and 15 feet comes to...................   .52
                                        ————
making............................$25.02
being the price for 715 feet at $35.00 per 1,000. If 8 feet is wanted, take twice what is given for 4 feet; if 6 feet, twice 3 feet; if 7 feet, take 3 and 4 feet—same way of dollars.

PRICE OF LUMBER

PER FOOT OF 1,000 FEET, BOARD MEASURE

No. feet	$ c. .25	$ c. .50	$ c. .75	$ c. 1.00	$ c. 1.25	$ c. 1.50	$ c. 2.00	$ c. 3.00
1	.00	.00	.00	.00	.00	.00	.00	.00
2	.00	.00	.00	.00	.00	.00	.00	.00
3	.00	.00	.00	.00	.00	.00	.00	.00
4	.00	.00	.00	.00	.00	.00	.00	.01
5	.00	.00	.00	.00	.00	.00	.01	.02
10	.00	.00	.00	.01	.01	.02	.02	.03
15	.00	.00	.01	.02	.02	.02	.02	.05
20	.01	.01	.02	.02	.03	.03	.04	.06
25	.01	.01	.02	.03	.03	.04	.05	.08
50	.01	.01	.04	.05	.06	.08	.10	.15
75	.02	.04	.06	.08	.09	.11	.15	.23
100	.03	.05	.08	.10	.13	.15	.20	.30
125	.03	.06	.09	.13	.16	.19	.25	.38
150	.04	.08	.11	.15	.19	.22	.30	.45
175	.04	.09	.13	.18	.22	.26	.35	.53
200	.05	.10	.15	.20	.25	.30	.40	.60
300	.08	.15	.23	.30	.38	.45	.60	.90
400	.10	.20	.30	.40	.50	.60	.80	1.20
500	.13	.25	.38	.50	.63	.75	1.00	1.50
600	.15	.30	.45	.60	.75	.90	1.20	1.80
700	.18	.35	.53	.70	.88	1.05	1.40	2.10
800	.20	.40	.60	.80	1.00	1.20	1.60	2.40
900	.23	.45	.68	.90	1.23	1.35	1.80	2.70
1000	.25	.50	.75	1.00	1.25	1.50	2.00	3.00
1500	.38	.75	1.13	1.50	1.88	2.25	3.00	4.50
2000	.50	1.00	1.50	2.00	2.50	3.00	4.00	6.00
2500	.63	1.25	1.88	2.50	3.13	3.75	5.00	7.50
3000	.75	1.35	2.25	3.00	3.75	4.50	6.00	9.00
4000	1.00	2.00	3.00	4 00	5.00	6.00	8.00	12.00
5000	1.25	2.50	3.75	5.00	6.25	7.50	10.00	15.00

PRICE OF LUMBER

PER FOOT OF 1,000 FEET, BOARD MEASURE

No feet	$ c. 4.00	$ c. 5.00	$ c. 6.00	$ c. 7.00	$ c. 8.00	$ c. 9.00	$ c. 10.00
1	.00	.01	.01	.01	.01	.01	.01
2	.00	.01	.01	.01	.02	.02	.02
3	.01	.02	.02	.02	.02	.03	.03
4	.02	.02	.02	.03	.03	.04	.04
5	.02	.03	.03	.04	.04	.05	.05
10	.04	.05	.06	.07	.08	.09	.10
15	.06	.08	.09	.11	.12	.14	.15
20	.08	.10	.12	.14	.16	.18	.20
25	.10	.13	.15	.18	.20	.23	.25
50	.20	.25	.30	.35	.40	.45	.50
75	.30	.38	.45	.54	.60	.68	.75
100	.40	.50	.60	.70	.80	.90	1.00
125	.50	.63	.75	.88	1.00	1.13	1.25
150	.60	.75	.90	1.05	1.20	1.35	1.50
175	.70	.88	1.05	1.23	1.40	1.58	1.75
200	.80	1.00	1.20	1.40	1.60	1.80	2.00
300	1.20	1.50	1.80	2.10	2.40	2.70	3.00
400	1.60	2.00	2.40	2.80	3.20	3.60	4.00
500	2.00	2.50	3.00	3.50	4.00	4.50	5.00
600	2.40	3.00	3.60	4.20	4.80	5.40	6.00
700	2.80	3.50	4.20	4.90	5.60	6.30	7.00
800	3.20	4.00	4.80	5.60	6.40	7.20	8.00
900	3.60	4.50	5.40	6.30	7.20	8.10	9.00
1000	4.00	5.00	6.00	7.00	8.00	9.00	10.00
1500	6.00	7.50	9.00	10.50	12.00	13.50	15.00
2000	8.00	10.00	12.00	14.00	16.00	18.00	20.00
2500	10.00	12.50	15.00	17.50	20.00	22.50	25.00
3000	12.00	15.00	18.00	21.00	24.00	27.00	30.00
4000	16.00	20.00	24.00	28.00	32.00	36.00	40.00
5000	20.00	25.00	30.00	35.00	40.00	45.00	50.00

PRICE OF LUMBER

PER FOOT OF 1,000 FEET, BOARD MEASURE

No. feet	$ c. 15.00	$ c. 20.00	$ c. 25.00	$ c. 30.00	$ c. 35.00	$ c. 40.00	$ c 50.00
1	.02	.02	.03	.03	.04	.04	.05
2	.03	.04	.05	.06	.07	.08	.10
3	.05	.06	.08	09	.11	.12	.15
4	.06	.08	.10	.12	.14	.16	.20
5	.08	.10	.13	.15	.18	.20	.25
10	.15	.20	.25	.30	.35	.40	.50
15	.23	.30	.38	.45	.52	.60	.75
20	.30	.40	.50	.60	.70	.80	1.00
25	.38	.50	.63	.75	.88	1.00	1.25
50	.75	1.00	1.25	1.50	1.75	2.00	2.50
75	1.13	1.50	1.88	2.25	2.63	3.00	3.75
100	1.50	2.00	2.50	3.00	3.50	4.00	5.00
125	1.88	2.50	3.13	3.75	4.38	5.00	6.25
150	2.25	3.00	3.75	4.50	5.25	6.00	7.50
175	2.63	3.50	4.38	5.25	6.13	7.00	8.75
200	3.00	4.00	5.00	6.00	7.00	8.00	10.00
300	4.50	6.00	7.50	9.00	10.50	12.00	15.00
400	6.00	8.00	10.00	12.00	14.00	16.00	20.00
500	7.50	10.00	12.50	15.00	17.50	20.00	25.00
600	9.00	12.00	15.00	18.50	21.00	24.00	30.00
700	10.50	14.00	17.50	21.00	24.50	28.00	35.00
800	12.00	16.00	20.00	24.00	28.00	32.00	40.00
900	13.50	18.00	22.50	27.00	31.50	36.00	45.00
1000	15.00	20.00	25.00	30.00	35.00	40.00	50.00
1500	22.50	30.00	37.50	45.00	52.50	60.00	75.00
2000	30.00	40.00	50.00	60.00	70.00	80.00	100.00
2500	37.50	50.00	62.50	75.00	87.50	100.00	125.00
3000	45.00	60.00	75.00	90.00	105.00	120.00	150.00
4000	60.00	80.00	100.00	120.00	140.00	160.00	200.00
5000	75.00	100.00	125.00	150.00	175.00	200.00	250.00

STAVE AND HEADING BOLTS

EXPLANATION OF RULE FOR TABLE.—Suppose a load to contain 25 feet at $2.75 per cord, look at 25 feet and under $2.75 opposite 25 you will find $2.15 the cost of 25 feet. If the price is wanted at $4.50 or $6.75 per cord, you first find price of the load at $4.00 or $6.00, then at 56 cts. or 75 cts., and add the two amounts together, so of other numbers.

SIMPLE RULE FOR MEASURING LOADS.—As per table, divide the price per cord by 32, the number of feet in a cord, *i.e.*, $6.00, the price per cord divided by 32, the number of feet in a cord gives you 19 cents per foot. When fractions occur, if over ½, add one; if less, nothing.

They are usually sold by the wagon load, at so much per cord, a cord being 8 feet long and 4 feet high—32 feet—width not taken into account. For Stave bolts the following timber is generally used in the Northern States. White Ash, Elm and Red Oak; it should be sound and free from knots and bark, and got out in proper shape, as per diagram.

HEADING BOLTS are generally made of sound Bass Wood, or Whitewood, timber either 18 inches long or 37 inches, and not less than 8 inches in diameter. If from 8 to 12 inches in diameter, leave them whole; if from 12 to 18 inches, halve them; if over 18 inches, quarter.

STAVE BOLTS are made 32 inches long.

STAVE AND HEADING BOLT TABLE

PRICE PER CORD

Ft.	.12½	.25	.37½	.50	.62½	.75	.87½	$1.00	$1.25
1	.00	.01	.01	.02	.02	.02	.03	.03	.04
2	.00	.02	.02	.03	.04	.05	.05	.06	.08
3	.01	.02	.04	.04	.06	.07	.08	.09	.12
4	.02	.03	.05	.06	.08	.09	.11	.12	.16
5	.02	.04	.06	.08	.10	.12	.14	.16	.20
6	.02	.05	.07	.09	.12	.14	.16	.19	.23
7	.03	.05	.08	.11	.14	.16	.19	.22	.27
8	.03	.06	.09	.12	.16	.19	.22	.25	.31
9	.04	.07	.11	.14	.18	.21	.25	.28	.35
10	.04	.08	.12	.16	.20	.23	.27	.31	.39
11	.04	.09	.13	.17	.21	.26	.30	.34	.43
12	.05	.09	.14	.19	.23	.28	.33	.37	.47
13	.05	.10	.15	.20	.25	.30	.36	.41	.50
14	.05	.11	.16	.22	.27	.33	.38	.44	.54
15	.06	.12	.18	.23	.29	.35	.41	.47	.58
16	.06	.13	.19	.25	.31	.38	.44	.50	.62
17	.07	.13	.20	.27	.33	.40	.46	.53	.66
18	.07	.14	.21	.28	.35	.42	.49	.56	.70
19	.07	.15	.22	.30	.37	.44	.52	.59	.74
20	.08	.16	.23	.31	.39	.47	.55	.62	.78
21	.08	.16	.25	.33	.41	.49	.57	.66	.82
22	.09	.17	.26	.34	.43	.52	.60	.69	.86
23	.09	.18	.27	.36	.45	.54	.63	.72	.90
24	.09	.19	.28	.37	.47	.56	.66	.75	.94
25	.10	.19	.29	.39	.49	.59	.68	.78	.98
26	.10	.20	.30	.41	.51	.61	.71	.81	1.00
27	.11	.21	.32	.42	.53	.63	.74	.84	1.04
28	.11	.22	.33	.44	.55	.66	.77	.87	1.08
29	.11	.22	.34	.45	.57	.68	.79	.91	1.13
30	.12	.23	.35	.47	.59	.70	.82	.94	1 17
31	.12	.24	.36	.48	.61	.73	.85	.97	1.21
32	.13	.25	.38	.50	.63	.75	.88	1.00	1.25

STAVE AND HEADING BOLT TABLE
PRICE PER CORD

Ft.	1.50	1.75	2.00	2.25	2.50	$2.75	$3.00	$3.25	$3.50
1	.05	.05	.06	.07	.08	.08	.09	.10	.11
2	.09	.11	.12	.14	.15	.17	.19	.21	.22
3	.14	.16	.19	.21	.23	.26	.28	.30	.32
4	.19	.22	.25	.28	.31	.34	.37	.41	.44
5	.23	.27	.31	.35	.39	.43	.47	.51	.55
6	.28	.33	.37	.42	.46	.51	.56	.61	.65
7	.33	.38	.44	.49	.55	.60	.66	.71	.77
8	.38	.43	.50	.56	.62	.69	.75	.81	.87
9	.42	.49	.56	.63	.70	.77	.84	.91	.98
10	.47	.55	.62	.70	.78	.85	.94	1.03	1.10
11	.52	.60	.69	.77	.86	.95	1.03	1.12	1.20
12	.56	.66	.75	.84	.94	1.03	1.12	1.21	1.30
13	.61	.71	.81	.91	1.01	1.11	1.22	1.32	1.42
14	.65	.77	.87	.98	1.09	1.20	1.31	1.42	1.53
15	.70	.82	.94	1.06	1.17	1.29	1.41	1.53	1.64
16	.75	.88	1.00	1.13	1.25	1.38	1.50	1.63	1.75
17	.80	.93	1.06	1.19	1.33	1.46	1.59	1.72	1.86
18	.84	.98	1.12	1.26	1.40	1.54	1.69	1.83	1.97
19	.89	1.04	1.19	1.34	1.49	1.63	1.78	1.93	2.08
20	.94	1.10	1.25	1.41	1.56	1.72	1.87	2.03	2.18
21	.98	1.15	1.31	1.47	1.64	1.80	1.97	2.13	2.30
22	1.03	1.20	1.37	1.54	1.71	1.89	2.06	2.23	2.40
23	1.08	1.26	1.44	1.62	1.80	1.98	2.16	2.34	2.52
24	1.12	1.31	1.50	1.69	1.87	2.06	2.25	2.44	2.62
25	1.17	1.37	1.56	1.75	1.95	2.15	2.34	2.53	2.73
26	1.22	1.42	1.62	1.82	2.03	2.23	2.44	2.64	2.85
27	1.27	1.48	1.69	1.90	2.11	2.32	2.53	2.74	2.95
28	1.31	1.53	1.75	1.97	2.19	2.41	2.62	2.84	3.06
29	1.36	1.59	1.81	2.03	2.26	2.45	2.72	2.94	3.17
30	1.41	1.64	1.87	2.10	2.34	2.57	2.81	3.04	3.28
31	1.45	1.70	1.94	2.18	2.42	2.67	2.91	3.15	3.39
32	1.50	1.75	2.00	2.25	2.50	2.75	3.00	3.25	3.50

STAVE AND HEADING BOLT TABLE

PRICE PER CORD

Ft.	$3.75	$4.00	$5.00	$6.00	$7.00	$8.00	$9.00	10.00
1	.11	.13	.16	.19	.22	.25	.28	.31
2	.24	.25	.31	.37	.44	.50	.56	.62
3	.35	.37	.47	.56	.66	.75	.84	.94
4	.43	.50	.62	.75	.87	1.00	1.12	1.25
5	.59	.62	.78	.94	1.09	1.25	1.42	1.56
6	.70	.75	.93	1.12	1.31	1.50	1.69	1.87
7	.82	.87	1.09	1.31	1.53	1.75	1.97	2.18
8	.94	1.00	1.25	1.50	1.75	2.00	2.25	2.50
9	1.05	1.12	1.41	1.69	1.97	2.25	2.53	2.81
10	1.17	1.25	1.56	1.87	2.19	2.50	2.81	3.12
11	1.29	1.37	1.72	2.06	2.40	2.75	3.09	3.44
12	1.40	1.50	1.87	2.25	2.62	3.00	3.34	3.75
13	1.52	1.62	1.93	2.43	2.84	3.25	3.65	4.06
14	1.64	1.75	2.19	2.62	3.06	3.50	3.94	4.37
15	1.75	1.87	2.34	2.81	3.28	3.75	4.22	4.69
16	1.88	2.00	2.50	3.00	3.50	4.00	4.50	5.00
17	1.99	2.12	2.67	3.19	3.72	4.25	4.79	5.31
18	2.11	2.25	2.81	3.37	3.94	4.50	5.06	5.62
19	2.22	2.37	2.97	3.56	4.16	4.75	5.34	5.94
20	2.34	2.50	3.12	3.75	4.37	5.00	5.62	6.25
21	2.46	2.62	3.28	3.94	4.59	5.25	5.90	6.56
22	2.58	2.75	3.44	4.12	4.81	5.50	6.18	6.87
23	2.70	2.87	3.59	4.31	5.03	5.75	6.46	7.19
24	2.81	3.00	3.75	4.50	5.25	6.00	6.75	7.50
25	2.93	3.12	3.91	4.69	5.47	6.25	7.03	7.81
26	3.05	3.25	4.06	4.87	5.69	6.50	7.31	8.12
27	3.16	3.37	4.22	5.06	5.90	6.75	7.59	8.44
28	3.28	3.50	4.37	5.25	6.12	7.00	7.87	8.75
29	3.40	3.62	4.53	5.43	6.34	7.25	8.15	9.06
30	3.51	3.75	4.68	5.62	6.56	7.50	8.43	9.37
31	3.64	3.87	4.84	5.81	6.78	7.75	8.71	9.69
32	3.75	4.00	5.00	6.00	7.00	8.00	9.00	10.00

RULES FOR CALCULATING SPEED OF SAWS, PULLEYS OR DRUMS

PROBLEM 1. The diameter of the driver being given, to find its number of revolutions.

RULE: Multiply the diameter of the driver by its number of revolutions, and divide the product by the diameter of the driven; the quotient will be the number of revolutions of the driven.

PROBLEM 2. The diameter and revolutions of the driver beng given, to find the diameter of the driven, that shall make any number of revolutions in the same time.

RULE: Multiply the diameter of the driver by its number of revolutions, and divide the product by the revolutions of the driven; the quotient will be its diameter.

PROBLEM 3. To ascertain the size of the driven.

RULE: Multiply the diameter of the driven by the number of revolutions you wish it to make, and divide the product by the revolutions of the driver; the quotient will be the size of the driven.
—Emerson, Smith & Co.

CAPACITY OF CIRCULAR SAW MILLS

TO THE HORSE POWER.—"How much lumber to each Horse Power will a Circular Saw Mill cut?" is often asked. A Horse Power is that which will raise 33,000 pounds one foot high per minute; 12 superficial feet of heating surface on a boiler, is supposed, under ordinary circumstances, to generate steam for one-horse power. In a large mill of 30-Horse Power capacity, each Horse Power ought to manufacture 1,000 feet of lumber; but in smaller mills, proportionately less. A 10-Horse Power ought to manufacture or saw 5,000 feet per 12 hours. Mills of larger power than 30 to 40-horse, ought, and generally do, over-run 1,000 feet to the horse power.

SIZE OF BOXES FOR DIFFERENT MEASURES

A box 24 inches long by 16 inches wide, and 28 inches deep, will contain a barrel (3 bushels).

A box 24 inches long by 16 inches wide, and 14 inches deep, will contain half a barrel

A box 16 inches square and 8 2-5 inches deep, will contain one bushel.

A box 16 inches by 8 2-5 inches wide, and 8 inches deep, will contain half a bushel.

A box 8 inches by 8 2-5 inches square, and 8 inches deep, will contain one peck.

A box 8 inches by 8 inches square, and 4 1-5 inches deep, will contain one gallon.

A box 7 inches by 4 inches square, and 4 4-5 inches deep, will contain half a gallon.

A box 4 inches by 4 inches square, and 4 1-5 inches deep, will contain one quart.

In purchasing anthracite coal 20 bushels are generally allowed for a ton.

TABLE OF SPEED OF CIRCULAR SAWS

Size of Saw.	Rev. per min.	Size of Saw.	Rev. per min
8 in	4,500	42 in	870
10 in	3,600	44 in	840
12 in	3,000	46 in	800
14 in	2,585	48 in	750
16 in	2,222	50 in	725
18 in	2,000	52 in	700
20 in	1,800	54 in	675
22 in	1,636	56 in	650
24 in	1,500	58 in	625
26 in	1,384	60 in	600
28 in	1,285	62 in	575
30 in	1,200	64 in	550
32 in	1,125	66 in	545
34 in	1,058	68 in	529
36 in	1,000	70 in	514
38 in	950	72 in	500
40 in	900	74 in	485
Shingle Machine Saws			1,400

NINE thousand feet per minute, that is nearly two miles per minute, for the rim of a circular saw to travel, may be laid down as a rule. For example, a saw 12 inches in diameter, three feet around the rim, 3,000 revolutions; 24 inches in diameter, or 6 feet around the rim, 1,500 revolutions; 3 feet in diameter, or 9 feet around the rim, 1,000 revolutions, etc. Of course it is understood that the rim of the saw will run a little faster than this reckoning, on account of the circumference being more than three times as large as the diameter. Shingle and some other saws, either riveted to a cast iron collar, or very thick at the centre and thin at the rim, may be run with safety at a greater speed.

POWER REQUIRED FOR CIRCULAR SAWS

To drive a 20 to 30 inch circular saw, 4 to 6 H. P.

" 32 to 40 " " 12 "

" 48 to 50 " " 15 "

" 50 to 62 " " 25 "

A VERY USEFUL TABLE

THE following table, computed from actual experience, will be found very useful in calculating the weight of loads, etc., or the weight of any of the articles in bulk. It shows the weight per cubic foot:

Cast Iron	450 lbs.	Common Soil, compact	124 lbs.
Water	62½ "	Clay, about	135 "
White Pine, seasoned, about	30 "	Clay, with stones	160 "
White Oak, seasoned, about	52 "	Marble	166 "
		Granite	169 "
Loose Earth	95 "	Brick	125 "

EBONY WOOD weighs eighty-three pounds to the cubic foot; lignum vitæ, the same; hickory, fifty-two pounds; birch, forty-five pounds; beech, forty; yellow pine, thirty-eight; white pine, twenty-five; cork, fifteen; and water, sixty-two.

FORTY feet of round, or 50 feet of hewn timber, one ton.

FORTY-TWO cubic feet one ton of shipping.

A CONVENIENT WOOD HOLDER

IT consists simply of a portion of a hollow log sawed off squarely, about one foot long and placed on one end for holding the wood while it is being split into small sticks. Such a contrivance saves labor, as it keeps the sticks erect, so that a workman may swing his axe freely; also saves time in picking up and adjusting the billets to be split. To prevent the numerous blows in one place from splitting such a holder, pin a half-round stick on the upper end, against which the axe may strike.

FENCE BOARD TABLE

SHOWING THE NUMBER OF FEET, BOARD MEASURE, RE-
QUIRED TO BUILD A FENCE FROM ONE TO FIVE
BOARDS HIGH, ¼ TO 1 MILE IN LENGTH

NO. BOARDS HIGH	1 MILE	½ MILE	¼ MILE
One	2,640 feet.	1,320 feet.	660 feet.
Two	5,280 "	2,640 "	1,320 "
Three	7,920 "	3,960 "	1,980 "
Four	10,560 "	5,280 "	2,640 "
Five	13,200 "	9,600 "	3,300 "

RAILWAY CROSS=TIES

NUMBER PER MILE, SINGLE TRACK

18 inches from centre to centre				3,520 ties.
21 " " "				3,017 "
24 " " "				2,640 "
27 " " "				2,348 "
30 " " "				2,113 "
33 " " "				1,921 "
36 " " "				1,761 "

GRADE PER MILE

THE following table will show the grade per mile as thus indicated:

An inclination of 1 foot in 10 is 528 feet per mile.

"	"	1	"	15 is 352	"	"
"	"	1	"	20 is 264	"	"
"	"	1	"	25 is 211	"	"
"	"	1	"	30 is 176	"	"
"	"	1	"	35 is 151	"	"
"	"	1	"	40 is 132	"	"
"	"	1	"	50 is 106	"	"
"	"	1	"	100 is 53	"	"
"	"	1	"	125 is 42	"	"

BRICKS

BRICKS may be estimated at 24 to a cubic foot, and five courses to one foot in height. But as bricks are not often of full size, the following allowances are made for each square foot of the surface, on the face of a wall, namely:

8 inch wall......................16 to a square foot.
12 "   " 24   "   "
16 "   " 32   "   "
20 "   " 40   "   "

CHIMNEYS

BRICKS, for chimneys, may be estimated for each foot in height, as follows:

Size of Chimney	Size of Flue	Number of Bricks to each foot in height
16 x 16	8 x 8	30
20 x 20	12 x 12	40
16 x 24	8 x 16	40
20 x 24	12 x 16	45

FRAMING TIMBER

IN a large class of houses, the following dimensions are sufficient, and are much used, namely:

Sills	7 x 8	Plates	3 x 6
Floor Timber	2 x 8	Rafters	4 x 5
Posts	4 x 6	Studding for partitions	2 x 3
Tie Beams	4 x 7	Furring	1 x 3
Studs	2 x 4		

SIZE OF NAILS

THE following table will show at a glance, the length of the various sizes, and the number of nails in a pound; they are rated 3-penny up to 20-penny.

Number	Length in inches	Nails per pound
3-penny	1	557
4-penny	1½	535
5-penny	1¾	282
6-penny	2	177
7-penny	2¼	141
8-penny	2½	101
10-penny	2¾	68
12-penny	3	54
20-penny	3½	34

FROM the foregoing table an estimate of quantity and suitable size for any job of work can easily be made.

———

COST OF VARIOUS STYLES OF FENCE, VARIED BY LOCALITY

Narrow Slat Picket Fence	$6.25	per rod.
Wide Slat Picket Fence	5.32	"
Common Stone Wall	3.00	"
Common Four-board Fence	2.00	"
Common Split Rail Fence	2.00	"
Virginia Split Rail Fence	1.50	"
Steel Barb Fence, four wires	.84	"

———

"VERY few of the great minds of this country have come from the city, or the cradle of the rich. The farm and the workshop have supplied by far the largest number of our eminent men."—*Dr. Hall.*

RELATIVE HARDNESS OF WOODS

TAKING shell bark as the highest standard of our forest trees, and calling that 100, other trees will compare as follows:

Shell-Bark Hickory	100	Yellow Oak	60
Pignut Hickory	96	White Elm	58
White Oak	84	Hard Maple	56
White Ash	77	Red Cedar	56
Dogwood	75	Wild Cherry	55
Scrub Oak	73	Yellow Pine	54
White Hazel	72	Chestnut	52
Apple Tree	70	Yellow Poplar	51
Red Oak	60	Butternut	43
White Beech	65	White Birch	43
Black Walnut	65	White Pine	30
Black Birch	62		

WEIGHTS OF CORD=WOOD

		Lbs.	Carbon
1 Cord of Hickory		4,468	100
"	Hard Maple	2,864	58
"	Beech	3,234	64
"	Ash	3,449	79
"	Birch	2,368	49
"	Pitch Pine	1,903	43
"	Canada Pine	1,870	42
"	Yellow Oak	2,920	61
"	White Oak	1,870	81
"	Red Oak	3,255	70
"	Lombardy Poplar	1,775	41

IN TANNING, four pounds of oak-bark make one pound of leather.

ROPES

TABLE, SHOWING WHAT WEIGHTS HEMP ROPE WILL
BEAR WITH SAFETY

CIRCUMFERENCE	POUNDS	CIRCUMFERENCE	POUNDS
1 inch.	200	3 inch.	1800
1¼ "	312.5	3¼ "	2112.5
1½ "	450	3½ "	2450
1¾ "	612.5	3¾ "	2812.5
2 "	800	4 "	3200
2¼ "	1012.5	5 "	5000
2½ "	1250	6 "	7200
2¾ "	1512.5		

NOTE.—A square inch of hemp fibres will support a weight of 9,200 pounds. The MAXIMUM strength of a good hemp rope is 6,400 pounds to the square inch. Its PRACTICAL value not more than one-half this strain. Before breaking, it stretches from one-fifth to one-seventh, and its diameter diminishes one-fourth to one-seventh. The strength of manilla is about one-half that of hemp. White ropes are one-third more durable. The strongest description of hemp rope is untarred, white three-strand rope; and the next in the scale of strength is the common three-strand, hawser-laid rope, tarred.

Wire rope is more than twice the strength of hemp of the same circumference.

Splicing a rope is estimated to weaken it one-eighth.

SHINGLES

SHINGLES are usually 16 inches long, and a bundle of shingles is 20 inches wide, and contains 24 courses in the thickness at each end; hence, a bundle of shingles will lay one course 80 feet long. When shingles are exposed 4 inches to the weather, 1,000 will cover 107 square feet; $4\frac{1}{2}$ inches, 120 square feet; 5 inches, 132 square feet; 6 inches, 160 square feet.

——————

DURABILITY OF SHINGLES

THE following table exhibits the average durability of shingles in exposed situations:

Rifted Pine Shingles.... from 20 to 35 years.
Sawed, clear from sap.....from 16 to 22 years.
Sawed, clear with sap..... from 4 to 17 years.
Cedar.... ....from 12 to 18 years.
Spruce.. ....from 7 to 11 years.

NOTE.—By soaking shingles in lime water, their durability is considerably increased.

——————

NUMBER OF SHINGLES required for a roof of any size; one which we think every mechanic and farmer should remember: First find the number of square inches in one side of the roof; cut off the right hand or unit figure, and the result will be the number of shingles required to cover both sides of the roof, laying five inches to the weather.   The ridge board provides for the double courses at the bottom. Illustration: Length of roof, 100 feet, width of one side, 30 feet—100 × 30 × 144 = 432,000. Cutting off the right hand figure we have 43,200 as the number of shingles required.

RIVED SHINGLES of clear pine are the best, not only because of the durability of the stuff in and of itself, but because the smooth cut of the drawing knife leaves the least possible roughness upon the surface for decay to take hold of. Next to these comes rived spruce and hemlock, which being far from as durable, may be placed near the peak of the roof, while the pine shingles are placed lower down, where the greater quantity of water passing over requires greater resistance to wear; sawed shingles have a rough surface, which holds water and causes rot.

GROWTH OF TREES

THE average growth of trees during 12 years, as determined by the Illinois Historical Society, when planted in belts and groves, is as follows:

White Maple.................1 ft. diam.	.....20 ft. high
Ash-leaf Maple............1 "	.....20 "
White Willow.............1½ "	.....40 "
Yellow Willow............1½ "	.....35 "
Blue and White Ash......10 in. diam.	.....20 "
Chestnut................10 "	.....20 "
Black Walnut..............10 "	.....20 "
Butternut................10 "	.....20 "
Elm......................10 "	.....20 "
Birch (varieties)..........10 "	.....20 "
Larch...................8 "	.....25 "

CORD WOOD ON AN ACRE

To estimate the quantity of cord wood on an acre of woodland requires experience. A person who has been engaged in clearing land and cutting wood could give a very close estimate at a general glance, but other persons would make the wildest guesses. An inexperienced person may proceed as follows: Measure out four square rods of ground; that is, thirty-three feet each way, and count the trees, averaging the cubic contents as near as possible of the trunks, and adding one-fourth of this for the limbs. Then, as 128 cubic feet make a cord, and the plot is one-fortieth of an acre, the result is easily reached. Fairly good timber land should yield a cord to every four square rods. A tree two feet in diameter and thirty feet high to the limbs, will make a cord of wood if it is growing in close timber, and the limbs are not heavy. If the limbs are large and spreading, such a tree will make 1¼ to 1½ cords. A tree one foot in diameter will make one-fourth as much as one twice the diameter. In estimating it is necessary to remember this fact.

The estimates given to the Department of Agriculture in different States, are as follows, so says the *Maine Farmer:* Several counties in Maine, 30 to 40 cords per acre. In New Hampshire, average yield 20 to 40 cords per acre. In Vermont, the forests yield 25 to 50 cords per acre. In Rhode Island, about 30 cords per acre. In Connecticut, sprout lands yield about 25 cords per acre every 25 years. In New York, 30 to 60 cords per acre. In Delaware, well set second growth wood lands yield 30 to 40 cords per acre. In Maryland, 30 to 40 cords. In Oregon, however, the yield of the evergreens and oaks is perfectly astounding, some counties estimated as high as 300 to 600 cords per acre.

HOW TO SAW VALUABLE TIMBER

ALL tough timber, when the logs are being sawed into lumber of any kind, whether scantling, boards or planks, will spring badly when a log is sawed in the usual manner, by commencing on one side and working toward the other. In order to avoid this, it is only necessary to saw off a slab or plank, alternately, from each side, finishing in the middle of the log. We will suppose, for example, that a log of tough timber is to be sawed into scantling of a uniform size. Let the sawing be done by working from one side of the log toward the other, and the end of the scantling will all be of the desired size, while at the middle some of them will measure one inch broader than at the ends. After the log has been spotted, saw off a slab from one side; then move the log over and cut a similar slab from the opposite side. Let calculations be made by measuring before the second slab is cut off, so that there will be just so many cuts, no more and no less, allowing for the kerf of every cut. If the log is to be cut into three-inch scantling, for example, saw a three-inch plank from each side, until there is a piece six and a quarter inches thick left at the middle. The kerf of the saw will remove about one-fourth of an inch. When a timber log is sawed in this way, the cuts will be of a uniform thickness from end to end. Now turn the log down, and saw the cuts the other way in the same manner, and the scantling will not only be straight, but of a uniform size from one end to the other, if the saw be started correctly.—*Selected.*

WELL=SEASONED FUEL

"THE best time to cut, haul and prepare wood for fuel is in the comparative leisure of winter, and where wood is used for fuel it should be thoroughly dried, as in its green and ordinary state it contains 25 per cent. of water; the heat to evaporate which is necessarily lost; therefore, the burning of green wood is greatly wasteful.

A log of unseasoned wood weighing, say 100 pounds, will weigh, when dry, only 66 pounds. What now has it lost? any combustible matter? anything that will warm your house or cook your food? No! it has lost 34 pounds of water. If about one-third the weight of green wood is water, then there are 1,443 pounds of water in a cord, this has to be made into steam before the wood can be burned. By drying the wood most of the water is expelled and there is little loss of heat in drying as it burns. Now, it costs about two dollars to work up a cord of wood for the stove after it is hauled to the wood pile, and it makes a difference that any one can calculate, whether a cord of wood burned green lasts twenty days, or burned dry lasts thirty days. A solid foot of green elm wood weighs 60 to 65 pounds, of which 30 to 35 pounds is sap or water. Beech wood loses one-eighth to one-fifth its weight in drying; oak, one-quarter to two-fifths. Therefore, get the winter's wood for fuel or kindlings and let it be seasoned as soon as possible, and not have a daily tussle with sissing firebrands and soggy wood."

SHAPE OF THE AXE

THE form of the edge of a chopping-axe should be determined by the purpose for which that tool is intended. When an axe is to be employed more for scoring timber than for chopping firewood, the form of the cutting edge should be nearly straight from one corner of the bit to the other, with the very corners rounded off, so that the axe will not stick badly in the timber. The object of having the axe nearly straight on the cutting edge, is to enable the chopper to score fully up to the line, without hacking the timber beyond the line. When the bit of the axe is what choppers term very circular, it is unfit to score timber with, as the most prominent part of the cutting edge will hack the surface of the timber a half-inch or more beyond the line. But by scoring with an axe that has nearly a straight edge, but few hacks may be seen after the timber has been hewed.

A good chopping axe should be rounded on the cutting edge and weigh from $3\frac{1}{2}$ to 5 pounds (some prefer lighter, others heavier), well hung on a tough, springy handle. (See illustration.)

WOODSMEN AND AXES

WE copy the following from the *Northwestern Lumberman:* "The styles of axes differ with nationalities. A Canadian chopper prefers a broad square blade, with the weight more in the blade than elsewhere, the handles being short and thick. A down-east logger, one from Maine, selects a long, narrow head, the blade in crescent shape, the heaviest part in the top of the head above the eye. New York cutters select a broad, crescent-shaped blade, the whole head rather short, and the weight balanced evenly above and below the eye, that is, where the handle goes through. A West backwoodsman selects a blade, the corners only rounded off, and the eye holding the weight of the axe. The American chopper, as a rule, selects a long, straight handle. The difference in handling is, that a down-easter takes hold with both hands at the extreme end, and throws his blows easily and gracefully, with a long sweep, over his shoulder. A Canuck chops from directly over his head, with the right hand well down on the handle to serve in jerking the blade out of the stick. A Westerner catches hold at the end of his handle, the hands about three inches apart, and delivers his blows rather directly from over the left shoulder.

In fact, an expert in the woods can tell the nationality or State a man has been reared in by seeing him hit one blow with an axe. It is, however, an interesting fact to know that a Yankee chopper, with his favorite axe and swinging cut, can, bodily strength being equal, do a fifth more work in the same time than any other cutter, and be far less fatigued. This, in a very large degree, will account for the great percentage of Maine men who will be found each year in the woods.

THE WEDGE is one of the mechanical powers—it has its place and is almost as indispensable among choppers as the axe. Its power to separate bodies from one another is perfectly wonderful. The power of the wedge increases as its length increases, or as the thickness of its back decreases.

BEECH TREE LEAVES.—The leaf of the beech tree, collected at autumn, in dry weather, form an admirable article for filling beds. The smell is grateful and wholesome; they do not harbor vermin; are very elastic, and may be replenished annually without cost.

SPLITTING RAILS

For split rails only straight grained timber
should be used. The logs being chosen, the
tools required are a maul, a few sharp-pointed
iron wedges, two axes, and a dozen wedges of
some tough, hard wood. The log to be split
should be first marked on the line of the split
with an axe driven by light blows of the maul.
Two iron wedges are then driven in by alternate
blows, and if the log is large, three will be needed. A single wedge may be buried in the center
of the log without splitting it, but by using two
at the same time an even seam will be opened.
Wooden wedges are then driven in the opening
on the side of the log, until it is split in halves
from end to end. If the timber is inclined to
run out and not split straight, drive an axe in
with the maul along the line where the timber
ought to split, and then an iron wedge along
this line; any "strings" which may remain can
be cut through with the axe. The half of the
log is then split in the manner shown in the
illustration in two quarters, commencing at one
end. The quarters are split somewhat differ-
ently. Instead of commencing at the end, the
sharp wedges are driven in the side, and the
central portion of the piece of timber is split off
first. The next layer is then taken, which is
split again into two parts, always driving the

wedges in the middle, and looking out for the running of the timber, and preventing it as already explained. The outside portion is then split into halves, and then into quarters, or into five rails if necessary.—*American Agriculturist.*

CHARCOAL

THE best quality of charcoal is made from oak, maple, beech and chestnut. Wood will furnish, when properly charred, about 20 per cent. of coal. A bushel of coal from pine weighs 29 pounds; a bushel of coal from hard wood weighs 30 pounds; 100 parts of oak make nearly 23 of charcoal; beech, 21; apple, 23.7; elm, 23; ash, 25; birch, 24; maple, 22.8; willow, 18; poplar, 20; red pine, 22.10; white pine, 23.

FELLING TIMBER

LARGE TREES of valuable timber are sometimes seriously injured by splitting when they fall, simply because those who cut them down do not know how to do it well. The engraving shows a large stump and tree, which was badly damaged in the felling, and another well cut and ready to fall. Almost every one who has been among the wood choppers, when they have felled large trees of tough timber, will recollect having seen the "butt logs" of many trees split, and the long splinters remaining on the stump,  which were pulled out of the tree. When a tree is designed for fire-wood, it is of no importance to fell it without damage; but when every foot in length is valued at $1.00 or more, it is of importance to know how to cut it down without damaging the butt log. If the wind does not blow, a large tree may be cut nearly off before it falls. The way is to leave a small strip on each side of the tree, while at the middle it is cut entirely through, as represented. When a tree leans, for example, to the north or south, it should always be cut to fall east or west, and always, if possible, at right angles to the way it leans. If cut to fall the way it leans, there is great danger that it will split at the butt.

If a large tree be cut nearly off on one side, it will fall on that side of the stump. For this reason, if a longer and deeper kerf be made on one side of a tree than the other, and the small one a few inches higher than

the large one, it will be easy to make a large tree fall in the direction desired. A tree may sometimes be sawed down quite as advantageously as felled with an axe, if a saw is in good order. (See illustration.) To facilitate starting a saw in the right direction, bore a hole horizontally into the tree about two inches deep, and drive in a wooden pin, on which the blade of the saw may rest, until the kerf is sufficiently deep to steady it. Decide where the tree is to be felled, and saw the side in that direction half off first, then saw the opposite side. Two broad and thin iron wedges should be driven after the saw into the kerf to prevent the saw being pinched so tightly that it cannot be worked nor drawn out. The ears on the end of a saw for felling timber should be secured with bolts, so that one may be removed, and the saw withdrawn, when it is difficult to knock out the wedges from the kerf.—*American Agriculturist.*

WEIGHT OF VARIOUS SUBSTANCES

AVOIRDUPOIS

1 cubic foot of bricks weighs 124 pounds; 1 do. clay, 250; 1 do. sand or loose earth, 95; 1 do. common soil, 124; 1 do. cork, 15; 1 do. marble, 161; 1 do. granite, 165; 1 do. cast iron, 450.55; 1 do. wrought iron, 486.65; 1 do. tin, 435; 1 do. white pine, 29.56; 1 do. elm, 34.9; 1 do. English oak, 60.04; 1 do. sea water, 64.3; 1 do. fresh water, 62.05; 1 do. air, .07529; 1 do. steam, .3889.

SAWING DOWN TREES

TRYING THE SOUNDNESS OF TIMBER

LET a person apply his ear to one end of the stick, while another, with hammer hits the other end with a gentle stroke. If the tree be sound and good, the stroke will be distinctly heard at the other end, though the tree should be a hundred feet or more in length.

HARDENING WOOD FOR PULLEYS

AFTER a wooden pulley is turned and rubbed smooth, boil it for about eight minutes in olive oil; then allow it to dry, when it will become almost as hard as copper.

CUBIC OR SOLID MEASURE

1728 cubic inches = 1 cubic foot
46656 cubic inches = 27 cubic feet = 1 cubic yard.
40 cubic feet of round timber = 1 ton.
50 cubic feet of hewn timber = 1 ton.
42 cubic feet of shipping timber = 1 ton.
16 cubic feet = 1 cord foot.
8 cord feet or 128 cubic feet = 1 cord of wood.

CUBIC WEIGHT TABLE

34 cubic feet of	Mahogany	weigh	1 ton.		
39 " "	Oak	"	1 "		
39 " "	Ash	"	1 "		
51 " "	Beech	"	1 "		
60 " "	Elm	"	1 "		
65 " "	Fir	"	1 "		
24 " "	Loose earth	"	1 "		

TO FIND THE WEIGHT OF TIMBER, BEAMS, POSTS AND JOISTS

Multiply length in feet by the breadth in inches and the depth in inches, and the products by one of the following factors:

For Elm, 2.92; Yellow Pine, 2.85; White Pine, 2.47; Dry Oak, 4.04.

TO GET A GEAR WHEEL OFF A SHAFT, upon which it has been shrunk, take it to the foundry and pour some melted iron around the hub, and it will heat and expand so quickly there will be no time for the shaft to get hot, and the gear will come off easily.

A SAWING MACHINE

A HARD TIMES HIRED MAN

This is the name given a device depicted and described not long ago by a Pennsylvania farmer in *The Rural New Yorker*. He says:

"The hard times compelled me to cut wood alone. The machine is easily understood. Three poles or rods make a frame for the saw to swing on. Another rod fastened to a bolt

at the top of the frame plays inside two pieces
of board. The saw is made fast to the lower
end of this rod, and then it will swing back and
forth as shown in the cut. You can have a
horse for the wood or drive stakes into the
ground with the top crossed, so as to hold the
logs.

"I can put up five cords in 10 hours with this
machine. Of course it takes some little time
to learn how to run the saw just right. In this
machine the stakes are 9 feet long for the sides
and 10 feet for the other. The pendulum on
which the saw is fastened is 8 feet long and has
holes bored in it so that it can be easily raised
or lowered. I use the 'horse' or stakes for sawing poles from 2 to 6 inches in diameter. For
sawing large logs I use a rolling platform like
that on buzz saws."

QUARTER SAWING HARDWOODS

There has been of late a revival of the discus-
sion of the most satisfactory and profitable
methods of quarter or rift-sawing lumber, but
these discussions seem, for the most part, to
have ignored some important considerations
that materially affect the question.

There are two objects to be gained in quarter-
sawing; one is, simply to present a durable
surface or to prevent undue warping and un-

even shrinkage; and the other is to develop the figure of the medullary rays, as in oak and sycamore. Let the former be all that is required and the process is a comparatively simple one, capable of being carried out with economy of material and labor.

For example, yellow pine edge-grained flooring is defined as presenting the edge of the groin to the surface at an angle of not less than 45 degrees with the annual rings of growth. This is usually done by cutting cants four or six inches thick from around the heart, and then ripping them into strips by means of the big saw itself, a gang saw, edger or some special machine. There is in this way but little waste, as the strips are all square-edged and pretty much the entire contents of the log can be used in some way or other.

In hardwoods proper a similar method can be used where it is not desirable to develop a figure; but when that is wished for, as in white oak, an entirely different method of procedure must be adopted. In order to get the characteristic figure in white oak, it is necessary to cut almost or quite directly toward the heart. By cutting cants only, two or three pieces from each would have a figure, but by frequently turning the log and cutting always nearly toward the center, a large number of figured pieces are secured. But the trouble with the method which produces the greatest number of pieces is that the boards are not square-edged, and there has to be a further treatment by the edger and waste of material, and some of the pieces are narrow. Furthermore, the process is a slow and expensive one.

Here comes in a chance for study of the conditions with which each concern is contending. It is a matter for careful calculation of costs and results. Much, also, depends upon the character of the material. In white oak this varies greatly, and large timber will produce a valuable material which will warrant the expenditure of time and labor which would not be justified in the smaller or coarser logs.—*The Timberman.*

———

The Builder and Wood-Worker remarks that many of the losses of fingers and hands sustained by operatives of small saws in factories could be obviated by following what it calls a golden rule: "Never put your hand back of a running saw." The temptation to reach back to remove or straighten something is natural and difficult to resist. But a man can never afford to get careless around a saw or any cutting tool. Precautions against accidents could be taken to much greater extent than they are.

EMERSON, SMITH & CO., BEAVER FALLS, PA.,

Say in their Book on Sawing: "The greatest wear of a saw is on the under sides of the teeth. File nearly to an edge (but not quite), leaving a short bevel of say 1-32 of an inch wide on the under side of the point. BUT IN NO INSTANCE FILE TO A FINE POINT AND THIN WIRE EDGE.

First.—Be sure that the saw hangs properly on the mandrel.

Second.—The saw must be in proper line with the carriage, and the carriage run true.

Third —The mandrel must be level, and run tight in the boxes.

Fourth.—Round off the saw so that all teeth will cut the same amount, and be sure that the VERY POINTS of the teeth are widest.

Fifth.—Do nearly all the filing on the under sides of the teeth, and see that they are WELL SPREAD at the points; file square and have them project alike on both sides of the saw.

Sixth.—If the saw heats in the center when the mandrel runs cool in the boxes, cool it off and line it into the log a little.

Seventh.—If the saw heats on the rim and not in the center, cool it off and line it out of the log a little.

IN FILING SOLID TOOTHED CIRCULAR SAWS keep the throats or roots of the teeth ROUND, or as the saws are when new. ANGLES OR SQUARE CORNERS filed at the roots of the teeth will almost invariably cause a saw to crack. THE BACK OR TOP OF THE TOOTH LEADS OR GUIDES THE SAW, and should be filed square across. The under sides of the teeth may be filed a little beveling on the teeth of saws that are bent alternately for the set so as to leave the outer corners of the cutting edge longest.

N. B.—There are many sawyers who are perfect masters of the business and will be successful with any good saw. Others not so well versed in the use of saws may find these directions useful.

HOW TO BE A SUCCESSFUL SAWYER

1st. Acquire sufficient knowledge of machinery to keep a mill in good repair.

2nd. See that both the machinery and saws are in good order.

3rd. It does not follow because one saw will work well that another will do the same on the same mandrel, or that even two saws will hang alike on the same mandrel, on the principle that no two clocks can be made to tick alike, no two saws can be made that will run alike.

4th. It is not well to file all of the teeth of circular saws from the same side of the saw, especially if each alternate tooth is bent for the set, but file one-half the teeth from each side of the saw, and of the teeth that are bent from you, so as to leave them on a slight bevel and the outer corner a little the longest.

5th. Never file any saw to too sharp or acute angles under the teeth, but on circular lines, as all saws are liable to crack from sharp corners.

6th. Keep your saw round so that each tooth will do its proportional part of the work, or if a reciprocating saw, keep the cutting points jointed on a straight line.

7th. The teeth of all saws wear narrowest at the extreme points; consequently. they must be kept spread so that they will be widest at the very points of the teeth; otherwise, saws will not work successfully.

8th. Teeth of all saws should be kept as near a uniform shape and distance apart as possible, in order to keep a circular saw in balance and in condition for business.—*Emerson, Smith & Co.*

EVERY 1-16 of an inch saved in the width of the kerf, saves one thousand feet of lumber in each 16,000 sawed; therefore, any mill cutting on an average 16,000 per day, will save 26,000 feet of lumber per month, being more than the entire expense of running the mill.

FILING THE TEETH OF SAWS AND THEIR CARE

THE great secret of putting any saw in the best possible order consists in filing the teeth in a given angle to cut rapidly; besides this, there should be just set enough in the teeth to cut a kerf as narrow as it can be made, and at the same time allow the blade to work freely without pinching. On the contrary, the kerf must not be so wide as to permit the blade to rattle when in motion. The very points of the teeth do the cutting; if one tooth is longer than those on either side of it, the short teeth do not cut although their points may be sharp. It is of the utmost importance to have saws that are used for cutting up large logs into lumber filed at such an angle as will insure the largest amount of work with the least expenditure of power.

SQUARING THE CIRCLE.—One-half of the diameter multiplied by the diameter, or seven-elevenths of the area of the circle, will give the area of an inscribed square. To find the side of an inscribed square, multiply one-fourth of the circumference by nine. When the circumference is given, to find the diameter, multiply by seven and divide by twenty-two. Eleven-fourteenths of the diameter gives exactly one-fourth of the circumference. The above solution is mathematically true.

CERTAIN TIMBERS of great durability, when framed together, act upon each other so as to produce mutual destruction. Experiments with cypress and walnut, and cypress and cedar, prove that they will rot each other while joined together, but on separation the rot will cease, and the timbers remain perfectly sound for a long period.

AS A RULE, hard, or close-grained woods are much more durable than soft, or open-grained ones. But there are some exceptions.

WEIGHT PER 1000 FT. OF SEASONED LUMBER

KIND	POUNDS
Ash	3,550
Cedar	2,925
Cypress	3,350
Beech	4,000
Cherry	3,720
Birch	2,950
Dogwood	3,930
Elm	3,220
Butternut	1,960
Chestnut	3,170
Maple	4,000
Oak	3,675
Poplar	3,056
Willow	2,780
Locust	3,800
Norway Spruce	2,670
Hemlock	2,350
Hickory	3,960
Walnut	3,690
Pitch Pine	4,150
Red Pine	3,075
Yellow Pine	2,890
White Pine	2,880

WEIGHTS OF WOOD

NAMES	No. of cubic feet in a ton
Oak, just felled	$32\frac{1}{2}$
Oak, seasoned	$48\frac{1}{4}$
Beech	42
Ash	$42\frac{1}{2}$
Apple Tree	$45\frac{1}{4}$
Plum Tree	$47\frac{1}{4}$
Maple	$47\frac{1}{2}$
Cherry Tree	50
Elm	$53\frac{1}{2}$
Walnut	$53\frac{1}{2}$
Red Pine	$54\frac{1}{2}$
Yellow Pine	55
White Pine	65
Chestnut	$59\frac{1}{4}$
Sycamore	$59\frac{1}{4}$
Willow	61
Poplar, common	93
Cedar	64

GREASE FOR BELTS.—Grease for belts, which renders them more adhesive and durable, can be obtained by mixing oil of resin with ten per cent talc. The grease is spread on the belt with a brush several times, or until the leather is so impregnated with it that it will not absorb any more. The operation is repeated after a period of some weeks, a smaller quantity of grease being used. The belts acquire more flexibility and power of resistance, and adhere better to the drums, and do not slip. The greasing is only required to be repeated every few months.

TRANSVERSE STRENGTH

TABLE, showing the transverse strength of timber 1 foot long and 1 inch square weight suspended from one end:

MATERIALS SEASONED	Breaking weight Lbs.	Weight borne safely Lbs.	Value for gen'l use Lbs.
White Oak.......	240	196	40
Chestnut.........	170	115	65
Yellow Pine......	150	100	62
White Pine......	135	95	64
Ash.............	175	105	77
Hickory.........	270	200	50

TABLE, showing transverse strength of iron, square bar, 2 inches by 1 foot long; weight suspended from one end:

MATERIAL	Breaking weight Lbs	Weight borne safely. Lbs.	Value	Value or gen'l use Lbs.
Cast Iron...	5781	4000	400	290

ROUND, 3 inches in diameter by 12 inches long; weight suspended from end:

MATERIAL	Breaking weight Lbs.	Weight borne safely. Lbs.	Value	Value for gen'l use Lbs.
Cast Iron...	12000	8000	240	175

NOTE.—The strength of a projecting beam is only *one-fourth* of what it would be if supported at both ends, and only *one-sixth* of what it would be if *fixed* at both ends. The former is to the latter as 2 to 3.

TO MEASURE THE HEIGHT OF A TREE
[See cut.]

WALK on level ground to a distance from the foot of the tree or object about equal to its presumed height. Lie on your back on the ground, stretched at full length. Let an assistant note on a perpendicular staff at your feet the exact point where your line of vision to the top of the object crosses the staff. Measure the height of this, B C, and your own height to your eyes, A B. Then as A B: B C:: A D: D E.

EXAMPLE.—The distance from my eyes to my feet is 5 feet 6 inches; from the ground to where the line of vision crosses the staff is four feet; from the point where my eyes were to the foot of the tree is 90 feet, what is the height of the tree?

As 5, 6: 4:: 90: about 65 feet, the height of the tree.—Ans.

ANOTHER WAY.—When a tree stands so that the length of its shadow can be measured, its height can be readily ascertained as follows: Set a stick upright—let it be perpendicular by the plumb line. Measure the length of the shadow of the stick. Then, as the length of the shadow of the stick is to the height of the stick, so is the length of the shadow of the tree to the height of the tree.

For example, if the height of the stick is four feet, and its shadow six feet in length, and the length of the shadow of the tree ninety feet, then 6: 4:: 90: (60) or sixty feet, the height of the tree. In other words, multiply the length of the shadow of the tree by the height of the stick, and divide by the length of the shadow of the stick.

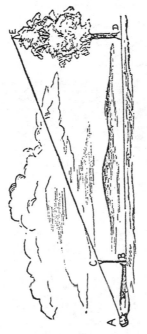

MEASURING THE HEIGHT OF A TREE

THE WOOD PILE

Wood cut during the three months that precede the first of the year is much more valuable than if cut the three months that succeed that time. The reason of this is, probably, because during the latter part of autumn, and the first part of winter, there is but little action in the sap of the tree, and therefore the wood is not filled with it, as it is after the sun runs higher and the days are longer. The strength of wood is proportionate to its weight. And as young trees grow more rapidly than old ones, they are more valuable as fuel. Round wood of oak or maple gives more heat than that which is so large as to be required to be split. Heart wood is heaviest, and the weight diminishes on proceeding outwards to the surface or upwards to the top of the tree, but less in old trees than in young growing ones.—*Selected.*

THE SHOP FOREMAN

It would seem, at first glance, that a shop foreman should be the best general workman in the establishment, and this is undoubtedly desirable if one can be found with the other qualifications necessary to a good foreman; but this is not often the case. Let us see what combination of qualities the best general workman must possess to make him eligible as a foreman. He must be a sober man who makes six days a week. He should have the confidence of his employers and the respect of the workmen. He should know how to manage as well as to command men. He must be able, in the shop at least, to entirely divest himself among the men of his old standard as a workman. He must be strictly impartial, and have the tact to find out the best way to get along with the men he has, and not those he would like to have. He must be able to plan ahead, have a good memory, a quick perception, be a rigid disciplinarian, and possess sound judgment; and because these qualifications are not often combined in the best workman is the reason why such a man is not always made foreman, and why the foreman is not always the best workman of the shop.—*Mechanical Engineer.*

LAND MEASURE

THE following table will assist farmers and others in making an accurate estimate of the amount of land in different fields:

10 Rods by	16 Rods,	1 Acre.			
8 "	20 "	1 "			
5 "	32 "	1 "		TO DRAW A RUSTED
4 "	40 "	1 "		NAIL OR SPIKE —First
5 Yds.	968 Yds.	1 "		drive it in a little,
10 "	484 "	1 "		which breaks the hold,
20 "	242 "	1 "		and then it may be
40 "	121 "	1 "		drawn out much eas-
220 Feet,	198 Feet,	1 "		ier.
110 "	396 "	1 "		
60 "	726 "	1 "		
120 "	363 "	1 "		
300 "	145.2 "	1 "		
400 "	108.9 "	1 "		

A WATERFALL is said to have a horse-power for every 33,000 lbs. of water passing a given point per minute for each foot of the fall. The following rule is given to compute the power of a waterfall, applied by James Watt:

RULE.—Divide the continued product of the width, the depth, the velocity of the water per minute, the height of the fall, and the weight of a cubic foot of water ($62\frac{1}{2}$ lbs.), by 33,000.

EXAMPLE.—The flume of a mill is 10 feet wide, the water is ten feet deep, the velocity is 100 feet per minute, and the fall 11 feet. What is the horse-power of the fall?

Operation. — $10 \times 3 \times 100 \times 11 \times 62\frac{1}{2} \div 33,000 = 62\frac{1}{2}$ H. P.

TABLE

EXHIBITING THE WEIGHT OF A LINEAL FOOT OF FLAT BAR IRON IN POUNDS

Breadth inches	Thickness in inches	Weight in pounds	Breadth inches	Thickness in inches	Weight in pounds	Breadth inches	Thickness in inches	Weight in pounds
1	¼	0.84	1⅞	1	6.33	2⅝	⅜	3.33
	½	1.69		1¼	7.92		½	4.43
	¾	2.53		1½	9.50		⅝	5.54
1⅛	¼	0.95	2	¼	1.70		¾	6.65
	½	1.90		⅜	2.53		⅞	7.76
	¾	2.85		½	3.38	2¾	⅛	1.16
1¼	¼	1.06	2⅛	⅛	0.90		¼	2.32
	½	2.11		¼	1.79		⅜	3.48
	¾	3.17		⅜	2.69		½	4.64
1⅜	¼	1.16		½	3.59		⅝	5.81
	½	2.32	2¼	⅛	0.95		¾	6.97
	¾	3.48		¼	1.90		⅞	8.13
1½	¼	1.26		⅜	2.85	2⅞	⅛	1.21
	½	2.53		½	3.80		¼	2.43
	¾	3.80	2⅜	⅛	1.00		⅜	3.64
1⅝	¼	1.37		¼	2.00		½	4.86
	½	2.74		⅜	3.01	3	⅛	1.27
	¾	4.12		½	4.01		¼	2.53
	1	5.49		⅝	5.02		⅜	3.80
	1¼	6.86		¾	6.02		½	5.07
	1½	8.24		⅞	7.02	3¼	¼	2.74
1¾	¼	1.48	2½	⅛	1.06		½	5.49
	½	2.96		¼	2.11		¾	8.23
	¾	4.43		⅜	3.17	3½	¼	2.95
	1	5.91		½	4.22		½	5.91
	1¼	7.39		⅝	5.28		¾	8.87
	1½	8.87		¾	6.33	4	¼	3.38
1⅞	¼	1.58		⅞	7.39		½	6.76
	½	3.17	2⅝	⅛	1.11		¾	10.14
	¾	4.75		¼	2.22			

TABLE

EXHIBITING THE WEIGHT OF A LINEAL FOOT OF ROUND
ROLLED IRON, FROM $\frac{1}{4}$ TO 4 INCHES DIAMETER

Diameter in ins.	Weight in lbs.	Diameter in ins.	Weight in lbs.	Diameter in ins.	Weight in lbs.	Diameter in ins.	Weight in lbs.
$\frac{1}{4}$	.165	$1\frac{1}{4}$	4.172	$2\frac{1}{4}$	13.440	$3\frac{1}{4}$	28.040
$\frac{3}{8}$	.373	$1\frac{3}{8}$	5.019	$2\frac{3}{8}$	14.975	$3\frac{3}{8}$	30.240
$\frac{1}{2}$	.663	$1\frac{1}{2}$	5.972	$2\frac{1}{2}$	16.688	$3\frac{1}{2}$	32.512
$\frac{5}{8}$	1.043	$1\frac{5}{8}$	7.010	$2\frac{5}{8}$	18.293	$3\frac{5}{8}$	34.686
$\frac{3}{4}$	1.493	$1\frac{3}{4}$	8.128	$2\frac{3}{4}$	20.076	$3\frac{3}{4}$	37.332
$\frac{7}{8}$	2.032	$1\frac{7}{8}$	9.333	$2\frac{7}{8}$	21.944	$3\frac{7}{8}$	39.864
1	2.654	2	10.616	3	23.888	4	42.464
$1\frac{1}{8}$	3.360	$2\frac{1}{8}$	11.988	$3\frac{1}{8}$	25.926		

EXAMPLE.—What is the weight of a bar of rolled iron, $1\frac{3}{4}$ inches diameter and 1 foot in length?

In column second find $1\frac{3}{4}$, and opposite to it is 8.128 lbs., which is 8 lbs. and $\frac{128}{1000}$ of a lb.; in the same way we may find the weight of any other diameter in the table.

TABLE

EXHIBITING THE WEIGHT OF A LINEAL FOOT OF SQUARE
ROLLED IRON, IN POUNDS, FROM $\frac{1}{4}$ TO 4 INCHES SQUARE

Size in ins.	Weight in lbs.	Size in ins.	Weight in lbs.	Size in ins.	Weight in lbs.	Size in ins.	Weight in lbs.
$\frac{1}{4}$	.211	$1\frac{1}{4}$	5.280	$2\frac{1}{4}$	17.112	$3\frac{1}{4}$	35.704
$\frac{3}{8}$	.475	$1\frac{3}{8}$	6.390	$2\frac{3}{8}$	19.066	$3\frac{3}{8}$	38.503
$\frac{1}{2}$	.845	$1\frac{1}{2}$	7.640	$2\frac{1}{2}$	21.120	$3\frac{1}{2}$	41.408
$\frac{5}{8}$	1.320	$1\frac{5}{8}$	8.926	$2\frac{5}{8}$	23.292	$3\frac{5}{8}$	44.418
$\frac{3}{4}$	1.901	$1\frac{3}{4}$	10.352	$2\frac{3}{4}$	25.560	$3\frac{3}{4}$	47.534
$\frac{7}{8}$	2.588	$1\frac{7}{8}$	11.883	$2\frac{7}{8}$	27.939	$3\frac{7}{8}$	50.756
1	3.380	2	13.520	3	30.416	4	54.084
$1\frac{1}{8}$	4.278	$2\frac{1}{8}$	15.263	$3\frac{1}{8}$	33.010		

NOTE.—The application of this table is the same as the preceding one.

EXPLANATION OF TABLE OF DAYS

ON the left you have the month, from any day of which to compute the number of days in any month. For example, you wish to know how many there are from the 20th of January to the 20th of August; following the line of January till you are under the month of August, gives you the number of days, 212, and so for other months.

SHINGLING, FLOORING AND PARTITIONING are usually measured by a square containing 100 square feet. 1,000 shingles are estimated to a square.

CEDAR, OAK AND CHESTNUT are the most durable woods in dry places.

ONE CUBIC FOOT of pure water, at 62° Fah., weighs 62.355 lbs.; at 212° Fah., only 56.640 lbs. A cylindrical foot of water, at 62° Fah., weighs 48.973 lbs. One ton of water is 35.90 cubic feet.

FACTS FOR BUILDERS

1,000 shingles, laid 4 inches to the weather, will cover 100 square feet of surface, and 5 lbs. of shingle nails will fasten them on.

One-fifth more siding and flooring is needed than the number of square feet of surface to be covered, because of the lap in the siding and matching.

1,000 laths will cover 70 yards of surface, and 11 lbs. of lath nails will nail them on. 8 bushels of good lime, 16 bushels of sand, and 1 bushel of hair, will make enough good mortar to plaster 100 square yards.

A cord of stone, three bushels of lime, and a cubic yard of sand will lay 100 cubic feet of wall.

TABLE

SHOWING THE NUMBER OF DAYS FROM ANY DAY IN ONE MONTH, TO THE SAME DAY IN ANY OTHER

FROM	JAN.	FEB.	MAR.	APRIL	MAY	JUNE	JULY	AUG.	SEPT.	OCT.	NOV.	DEC.
January.........	365	31	59	90	120	151	181	212	243	273	304	334
February........	334	365	28	59	89	120	150	181	212	242	273	303
March..........	306	337	365	31	61	92	122	153	184	214	245	275
April...........	275	306	334	365	30	61	91	122	153	183	214	244
May............	245	276	304	335	365	31	61	92	123	153	184	214
June...........	214	245	273	304	334	365	30	61	92	122	153	183
July...........	184	215	243	274	304	335	365	31	62	92	123	153
August.........	153	184	212	243	273	304	334	365	31	61	92	122
September......	122	153	181	212	242	273	304	334	365	30	61	91
October........	92	123	151	182	212	243	273	304	335	365	31	61
November......	61	92	120	151	181	212	242	273	304	334	365	30
December......	31	62	90	121	151	182	212	243	274	304	335	365

TABLE OF ELASTICITY AND STRENGTH OF VARIOUS KINDS OF TIMBER

NAME	Val. of E.	Val. of S.
English Oak	105.	1,672
Canadian Oak	155.5	1,706
Ash	119.	2,026
Beech	98.	1,556
Elm	50.64	1,013
Pitch Pine	88.68	1,632
Red Pine	133.	1,341
N. England Fir	158.5	1,102
Larch	76.	900
Norway Spruce	105.47	1,474

SHRINKAGE IN DIMENSIONS OF TIMBER BY SEASONING

WOODS	Inches
Pitch Pine	10 to 9¾
White Pine	12 to 11⅞
Pitch Pine, So.	18¾ to 18¼
Yellow Pine	18 to 17⅞
Spruce	8½ to 8⅜
Cedar, Canada	14 to 13¼
Elm	11 to 10¾
Oak	12 to 11⅝

THE WEIGHT of an ordinary lathed and plastered ceiling is about 10 pounds per square foot, and that of an ordinary floor of 1¼ inch boards, together with the usual 3 x 12 inch joist, 15 inches apart from center to center, is from 10 to 12 pounds per square foot; in preliminary calculations it is well to take the two together as 25 pounds per square foot.

BOARDS OF OAK OR PINE, nailed together by from 4 to 16 ten-penny cut nails, and then pulled apart in a direction lengthwise of the boards, and across the nails, tending to break the latter in two by a shearing action, averaged about 300 to 400 pounds per nail to separate them; the result of many trials.

THE CARE OF GRINDSTONES

THE exposure of the stone to the sun has a tendency to harden it. And if one part be left in the water habitually it will grow soft, and wear away faster than the other. If the trough is put upon movable supports in the frame, it can be adjusted to the stone without much loss of time. Or allow the water to drip from a water-pot, an old white-lead keg will answer, fixed above the stone. Always clean off all greasy or rusty tools before sharpening, as grease chokes up the grit; and always keep the stone perfectly round by razeeing it off when necessary.

To FACE AN OIL STONE put it into your pocket, if small, and carry it to some place where they cast iron, and rub it on a flat casting just come out of the sand. You can face it in ten minutes—use water on the iron.

POWER AND CAPACITY OF SAW MILLS

As a rule it is admitted by mill men that for 10,000 feet per day about 20 horse-power is required; for 20,000 feet, 30 horse-power; for 30,000 feet, 40 horse-power.

GOOD machinery is a necessity in the saw-mill, in the planing-mill, and in all wood-working establishments.

STONE WALL TABLE

EXPLANATION

FIND the length in the left, and the thickness in the right hand column; then follow down the column under the height, until you come to the line opposite the length and thickness, and you have the amount of feet required; then by adding or subtracting, you have the amount of any length, height or thickness desired. Inches under six in the whole amount, not mentioned—over six, called a foot.

STONE MASONRY is usually measured by the cubic foot, cubic yard or perch; a yard of stone wall is three feet long, three feet wide and 15 inches thick. A perch is $16\frac{1}{2}$ feet wide and 1 foot deep.

A CORD OF STONE, three bushels of lime, and a cubic yard of sand will lay one hundred cubic feet of wall.

SAND IS ESTIMATED by the load; a load containing from nineteen to twenty bushels. This is sufficient for about two casks of lime, therefore we may estimate one cask of lime to ten bushels of sand.

STONE WALL MEASURE

Length	HEIGHT IN FEET										Thickness
	1	2	3	4	5	6	7	8	9	10	
1	1	2	3	4	5	6	7	8	9	10	12
2	2	4	6	9	11	13	15	17	20	22	13
3	4	7	10	14	17	21	24	28	32	35	14
4	5	10	15	20	25	30	35	40	45	50	15
5	7	13	20	27	33	40	47	53	60	67	16
6	8	17	26	34	42	51	60	68	77	85	17
7	10	21	32	42	52	63	73	84	94	105	18
8	13	25	38	51	63	76	89	101	114	127	19
9	15	30	45	60	75	90	105	120	135	152	20
10	17	35	52	70	87	105	122	140	157	175	21
11	20	40	60	81	101	121	141	161	181	202	22
12	23	46	69	92	115	138	161	184	207	230	23
13	26	52	78	104	130	156	182	208	234	260	24

SUPPLIES FOR LUMBERING CREWS AND HORSES IN THE WOODS

THE following table will be found convenient as to the quantity and quality of supplies necessary for a lumberman's outfit in the woods for men and horses, of course varied by locality. Being the result of long experience in the business, it may be useful to many persons as a basis to make calculations for a lumbering crew.

50 lbs. of oats for each span of horses per day.

40 lbs. of hay for each span of horses per day.

As the work is severe, teams require to be well fed.

Quantity of Flour used by each man per day, 1.80

"	Beef	"	"	"	0.80
"	Pork	"	"	"	1.20
"	Potatoes	"	"	"	.45
"	Beans	"	"	"	.32
"	Onions	"	"	"	.12
"	Salt Fish	"	"	"	.12

Sugar and Molasses not always allowed. ——

Total daily consumption for each man, 4.81

Quantity of Tea for each man, per month, 1½ lbs.

" Coffee " " " 2 lbs.

———

To CURE SCRATCHES ON HORSES.—Wash their legs with warm soap suds, and then with beef brine. Two applications will cure the worst case.

STRENGTH OF ICE

ICE 2 inches thick will bear men on foot.
   "   4 inches thick will bear men on horseback.
   "   6 inches thick will bear cattle and teams
          with light loads.
   "   8 inches thick will bear teams with heavy
          loads.
   " 10 inches thick will sustain a pressure of
          1,000 pounds per square foot.

This supposes the ice to be sound through its whole thickness, without "snow-ice."

STAVES, ETC., COMPARED WITH BARRELS

In loading vessels, etc., with lumber, the following calculations may be relied on:

1,000 Barrel staves will require the room of 15 barrels.

1,000 Hogshead staves will require the room of 20 barrels.

1,000 Pipe staves will require the room of 30 barrels.

1,000 Feet of Boards will require the room of 20 barrels.

400 feet of Boards are rated at a ton.

TIMBER MEASURE is essential to the correct calculation of the cost of all wooden structures; it is constantly used by carpenters, joiners, etc., and is requisite to form estimates about their work.

REMOVING RUST FROM SAWS

PROCURE at some drug store a piece of pumice stone as large as a hen's egg, grind one side flat on a grind-stone, then scour off the rust with the pumice stone and soapsuds. Cover the surface with lard in which there is no salt.

ANOTHER.—Immerse the articles in kerosene oil and let them remain for some time, the rust will become so much loosened as to come off very easily.

WATER-PROOF LEATHER PRE-SERVATIVE

THIS is said to have been in use among New England fishermen for 100 years, when it was published in an almanac for 1794. "Take one pint boiled linseed oil, half a pound mutton suet, six ounces clean bees-wax, and four ounces resin; melt and mix over a fire, and apply while warm, but not hot enough to burn the leather. Lay it on plentifully with a brush, and warm it in.

A SUPERIOR LINIMENT

THE *Western Rural* says, that one of the very best liniments ever made, for man or beast, is composed of equal parts of laudanum, alcohol and oil of wormwood; its effect is almost magical.

CURE FOR SORE BACKS OF HORSES.—The best method of curing sore backs is to dissolve ½ an oz. blue vitriol in a pint of water, and daub the injured parts with it four or five times a day.

SAW MILL MEN

Saw mill men must remember that the most prominent defect that lowers the grade of lumber on inspection is bad manufacture. Of course this defect can be avoided, but it is one which often costs a man more than his profits.

PILING LUMBER

Lumber should not be allowed to depreciate for lack of proper care in piling. Piles should be built so that the front cross-piece shall be higher than the back, and each in succession be overlapped or laid out a trifle beyond the previous one. A pile twenty feet wide should incline outward from base to top at least eighteen or twenty-four inches, which will prevent storms from beating in, or snow from resting to melt and form ice. The sides of the pile should be carried up plump, each cross-piece directly on top of another, so that the weight shall rest solidly on each, and on the foundation timber. If the courses are placed a little forward or back of the previous one the weight above will twist, warp and perhaps break the lumber. Piles should never be placed less than three feet apart, and boards in the pile should alway be laid with from two to four inches of space between them.

TREATMENT OF LEATHER BELTS

ALL leather belts, especially those which are used in flour mills and wood-working establishments, are more or less subjected to dust, and no matter however soft and pliable a belt may be in the first instance, it is only a question of time when this fine dust which is constantly settling upon it will effectually suck out all the oil and render it hard and dry, and if the flesh side is run next to the pulley and the pulley of small diameter, fine cracks will appear upon the opposite side, crosswise of a depth corresponding to the state and condition of the belt, and these cracks frequently penetrate deep enough to materially impair its strength, and this is one strong reason, if nothing more, why the grain side of a belt should always run next to the pulley.

TO REMOVE WOOD from a file or rasp, dip the instrument in hot water, to swell the wood; it is then removed by a hand brush; the warmth evaporates the moisture.

HOW TO TREAT FROST BITES

A Doctor in Kansas City Star

DURING the past two days I have treated several people for frozen hands and feet. In one or two cases I have found it very difficult to treat them on account of their plunging their

frozen members in hot water or holding them
in close proximity to a red-hot stove.   The
best possible way to draw out the cold from
frozen parts is to plunge them into ice or snow-
water containing a liberal supply of saltpetre
or common salt, and the submitting to a vigorous rubbing with a coarse towel or slapping
with the hands to restore circulation.   In many
cases amputation has been found necessary
where the patient has foolishly applied hot
water.

REMEDIES FOR BURNS AND SCALDS

Every family should have a preparation of flax-
seed oil, chalk and vinegar, about the consistency
of thick paint, constantly on hand for burns and
scalds.   The best application in cases of burns
and scalds is a mixture of one part of carbolic
acid to 8 parts of olive oil.   Lint or linen rags
are to be saturated in the lotion, and spread
smoothly over the burned part, which should
then be covered with oiled silk or gutta-percha
tissue to exclude air.

CONVENIENT WOOD HOLDER

CARELESS PILING

IT is easier to make money than to take care of it. This is especially true in the lumber business. Much lumber is ruined in piling, not only from the sticks not being directly over each other, or with the slant of the pile, but from the ends being exposed. It is very common to see the sticks at each end back an inch or more. This allows the ends to dry quicker, and naturally must check the end, which is much worse in broad boards and in hardwoods. This is not all, The ends being overhung allow all the moisture to penetrate the pile. Stain and mildew are not pleasant to a customer buying clear finishing lumber.

CARE OF LUMBER

THE manufacture of lumber requires skill, but the taking care of, and properly assorting lumber, requires stricter attention. A glance into almost every lumber yard will demonstrate this by summing up the amount of waste arising in a dozen ways. Mill men lose as much by not properly assorting as in waste from bad piling, handling, etc. Every piece of lumber ought to be rigidly inspected as it comes from the trimmer. It is common among many mills to have much culling done from the stack in shipping or local trade, which necessitates extra handling and piling. In such yards the purchaser invariably discriminates more closely and wants only the best stuff, which is natural where a lot must be thrown aside that does not come up to the requirement. If mill men would restrict their output, and do it right from the saw, they would obtain better prices for their clear lumber and class the lower grades to suit the demand. It is an extremely difficult matter to deviate from this course with the local trade after adopting the slip-shod, pick-as-you-please, or log-run method. The merchant who does not grade his goods, but lumps them together, does not succeed.—*Southern Lumberman.*

————

A MANUFACTURER of wooden specialties asks if we ever noticed the difference between the length of a stick and a shaving taken from the same. He measured three shavings the other day. The stick from which he took them was 2 feet 10 inches long. The shavings were taken from the full length. The first shaving was .002 inch thick and shrunk a little over $2\frac{1}{4}$ inches. The next shaving was .005 inch thick and lacked nearly $1\frac{3}{4}$ inches of being as long as the stick. The third shaving was .012 inch thick and fell short $1\frac{1}{8}$ inches.—*Indianapolis Wood-Worker.*

FOR PULLING STUMPS

A DEVICE WHICH CAN BE RELIED UPON IN ALL
CIRCUMSTANCES

CUT a good strong pole about twenty feet long of white ash. Trim and peel it nicely, hitch a strong rope to the top—a chain will do, but it is heavier to handle. Set the pole against the stump to be pulled, letting the lower end rest between two roots. Then put a strong chain around the top of the stump, passing it around the pole. A team hitched to the rope will pull out most any stump. Place the

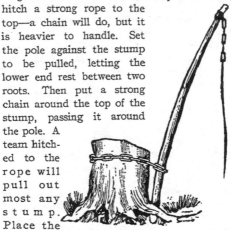

pole close to the stump and cut the roots opposite the pole. Two men can best do the work, one to tend the horse, the other to cut roots as the stump is being turned up.—*Farm and Home.*

MILL DAMS

WHEN building a dam, you should select the most suitable place. If you can, place it across the stream near a rocky bluff, so that the ends of the dam may run into the bluff. This will prevent the water running by at the ends of the dam. Build your dam strong. If this is not done, they are breaking up often, causing ruinous expense in money and loss of time.

PILE DRIVER.—In sandy soil, the greatest force of a pile-driver will not drive a pile over 15 feet.

MELTED SNOW produces from $\frac{1}{4}$ to $\frac{1}{8}$ of its bulk in water.

A FALL OF ONE INCH in a mile will produce a current in rivers. An inclination of three inches per mile in a straight, smooth channel will give a velocity of three miles per hour, while three feet per mile would produce a torrent.

STEEL, when hardened, decreases in specific gravity, contracts in length, and increases in diameter.

THE VALUE of a ton of pure Gold is $602,-799.21. $1,000,000 gold coin weigh 3,685.8 lbs. avoirdupois. The value of a ton of Silver is $37,704.84. $1,000,000 silver coin weigh 58,-929.9 lbs. avoirdupois.

TABLE SHOWING

THE DAY'S LENGTH, AT INTERVALS OF A WEEK
FOR THE YEAR

		Hours	Min.			Hours	Min.
Jan'y	1	9	9	July	1	15	13
	8	9	16		8	15	07
	15	9	26		15	14	58
	22	9	38		22	14	46
	29	9	52		29	14	33
Feb'y	4	10	09	Aug.	5	14	17
	12	10	26		12	14	01
	19	10	45		19	13	43
	26	11	04		26	13	24
March	4	11	24	Sept.	2	13	06
	11	11	44		9	12	46
	18	12	04		16	12	26
	25	12	24		23	12	06
April	1	12	44		30	11	47
	8	13	04	Oct.	7	11	27
	15	13	23		14	11	07
	22	13	42		21	10	48
	29	13	59		28	10	29
May	6	14	17	Nov.	4	10	11
	13	14	33		11	9	56
	20	14	47		18	9	40
	27	14	58		25	9	27
June	3	15	07	Dec.	2	9	17
	10	15	14		9	9	10
	17	15	17		16	9	06
	24	15	16		23	9	05
					30	9	08

SPIRITS OF TURPENTINE

THIS is one of the most valuable articles in
a family, and when it has once obtained a foot-
hold in the house it is really a necessity and
could be ill dispensed with. Its medicinal
qualities are very numerous; for burns it is a
quick application and gives immediate relief,
for blisters on the hands it is of priceless value,
searing down the skin and preventing soreness;
for corns on the toes it is useful, and good for
rheumatism and sore throats, and it is the
quickest remedy for convulsions or fits. Then
it is a sure preventive against moths; by just
dropping a trifle in the bottom of drawers,
chest and cupboards, it will render the gar-
ments secure from them during the summer.
It will keep ants and bugs from closets and
storerooms, by putting a few drops in the corners and upon the shelves; it is sure destruction to bedbugs and will effectually drive them
away from their haunts, if thoroughly applied
to the joints of the bedstead in the spring cleaning time, and injures neither furniture nor
clothing. Its pungency is retained for a long
time, and no family ought to be entirely out of
a supply at any time of the year.—*Practica
Farmer.*

POINTERS ABOUT STEAM BOILERS

According to *Steam* the requirements of a perfect steam boiler are:

1. The best materials sanctioned by use, simple in construction, perfect in workmanship, durable in use, and not likely to require early repairs.

2. A mud drum, to receive all impurities deposited from the water, in a place removed from the action of the fire.

3. A steam and water capacity sufficient to prevent any fluctuation in pressure or water level.

4. A large water surface for the disengagement of the steam from the water, in order to prevent foaming.

5. Constant and thorough circulation of water throughout the boiler, so as to maintain all parts at one temperature.

6. The water space divided into sections, so arranged that should any section give out no general explosion can occur, and the destructive effects will be confined to the simple escape of the contents; with large and free passages between the different sections to equalize the water line and pressure in all.

7. A great excess of strength over any legitimate strain; so constructed as not to be liable to be strained by unequal expansion, and, if possible, no joints exposed to the direct action of the fire.

8. A combustion chamber so arranged that the combustion of the gases commenced in the furnace may be completed before the escape to the chimney.

9. The heating surface, as nearly as possible, at right angles to the current of heated gases, and so as to break up the currents and extract the entire available heat therefrom.

10. All parts readily accessible for cleaning and repairs. This is a point of the greatest importance as regards safety and economy.

11. Proportioned to the work to be done, and capable of working to its full rated capacity with the highest economy.

12. The very best gauges, safety-valves and other fixtures.

THE SCIENTIFIC MACHINIST

I HAVE been troubled with the boxes on my crank pin getting warm and cutting away very rapidly. I could not locate the cause for some time. I had put my engine in perfect line, and still the trouble kept on. I was on the point of using some strong engineer's language, when I thought perhaps the trouble was in the strap of the crank pin boxes. So I took it off and again filled up my boxes and put them back, but instead of screwing the nuts tight before driving the key in, I inserted the bolt and drove the key down hard, and then tightened up the nuts. Then, loosening up the key, I drove it to the proper place. I have not been troubled with hot boxes since. This way you put your boxes in perfect position before you have made your strap fast.

———

PROPER TIME FOR CUTTING TIMBER

IF oak, hickory or chestnut timber be felled in August, in the second running of the sap, and barked, it will season perfectly, even a large tree; and the twigs will remain sound for years. Whereas that cut in winter and remaining till next fall, will be completely sap-rotten, and unfit for any purpose, almost. The body of the oak split into rails, will not last more than ten or twelve years. Chestnut will last longer, but no comparison to that cut in August. Hickory cut in August is not subject to be worm-eaten, and will last a long time for fencing. The tops of the trees cut in this month are more valuable for fuel than those cut in winter.

For resinous timber, such as pine, larch, etc., the proper time for cutting is during the months of June, July or August, as the pores of the wood will be filled with resin, which serves to increase the strength and durability of their timber.

HOW TO MEASURE WOOD PILE

To ascertain the number of cords of wood in a pile, multiply together the length, breadth and height, and divide by 128.

HOW TO MEASURE TIMBER

To ascertain the number of cubic feet in round timber, find the average circumference by adding the circumference of the larger and smaller ends and dividing by 2; multiply the square of one-fourth of this average circumference by the length in feet; the result gives four-fifths of the real contents in cubic feet; one-fifth being customarily allowed to the purchaser for waste in sawing.

To measure contents of square timber, multiply the width by the thickness in inches; this product by the length in feet, and divide by 12; result gives feet.

HOW TO MEASURE LUMBER

To measure boards, multiply length in feet by breadth in inches, and divide by 12 for inch boards; the quotient gives contents in feet. For boards $1\frac{1}{2}$ inches thick, add one-quarter to quotient; if $1\frac{1}{2}$, add one-half; if 2 inches, divide by 6 instead of 12; if 3 inches, divide by 4; if 4 inches, divide by 3; if 6 inches, divide by 2.

COMPARATIVE WEIGHT OF WOODS AND PERCENTAGE OF CARBON

	Pounds.	Carbon.
1 cord of Hickory	4,468	100
1 cord of White Oak	3,870	81
1 cord of Ash	3,449	79
1 cord of Red Oak	3,255	70
1 cord of Beech	3,234	64
1 cord of Yellow Oak	2,920	61
1 cord of Hard Maple	2,864	58
1 cord of Birch	2,368	49
1 cord of Pitch Pine	1,903	43
1 cord of Canada Pine	1,870	42
1 cord of Lombardy Poplar	1,775	41

THE DIFFERENT WOODS for charcoal may be estimated as to value by this rule. Of the oaks 100 parts will yield 23 parts charcoal, beech 21, apple, elm and white pine 23, birch 24, maple 22, willow 18, poplar 20, hard pine 22½. The charcoal used for gun-powder is made from willow and alder.

ALTHOUGH a lumber scribe but seldom drops into poetry, a four-line stanza may briefly depict the situation:

"O, woodman, cut that tree,
    Leave not a single bough;
    It will put five dollars in my inside pocket,
    Then why not cut it now."

HUMAN STRENGTH

An average strong man will, for a short period, exert a force with a

Drawing knife	equal to 100	lbs.
An auger, both hands	" 100	"
A screw driver, 1 hand	" 84	"
A bench vice, handle	" 72	"
A chisel, vertical pressure	" 72	"
A windlass	" 60	"
Pincers, compression	" 50	"
A hand-plane	" 50	"
A hand-saw	" 36	"
A thumb-vice	" 45	"
A brace-bit, revolving	" 16	"

THE HORSE

The strength of a horse is equivalent to five men.

A *draught horse* can draw 1,600 lbs., 23 miles a day on a level road, weight of carriage included.

The average weight of horses is 1,000 lbs. each.

A horse will carry 250 lbs., 25 miles a day of 8 hours.

He occupies in a stall a front of 4½ feet, and a depth of 10 feet.

ORIGIN OF THE WORD LUMBER

THE word "lumber," which has an essentially American origin as applied to manufactures of timber, was first used in Boston, in an official way, in 1663. It is a most comprehensive word, and other countries have no expression for it that covers the ground so completely. In Great Britain, for instance, each item of lumber has its name, as with us; but, if they were speaking of manufactures of wood as a whole, about the only term which they have that covers the case is "wood goods," which is an awkward expression at best. The word lumber was coined in Boston. A recent writer in the Boston *Journal* states that the word has not had full justice accorded to it. From 1630 for nearly one hundred years Boston was the chief lumber market of the world, and that industry was one of the principal foundations of Boston's wealth. Other Boston staples were fish and leather, but in magnitude of transactions lumber was in the lead. The site of the old state house, known as market place, was formerly a lumber yard. The men of Boston got to calling sawn timber lumber, because the ships that brought that article of commerce to Boston used to lumber up the wharves and streets with their product. In 1663 the police regulations of Boston provided that the wharves and all the streets "that butt upon the water" must be kept free from all "lumber and other goods." Boston lumber carried in Boston ships went to all parts of the world and laid the foundation for Boston wealth. It is said that the first cargo returned by the Pilgrim Fathers to England was a cargo of pipe staves, and for the reason that Europe could not produce as good an article, it was a profitable venture, netting the shippers five hundred pounds. In that industry the Puritans were satisfied that all Europe could not rival them. The term lumber included masts, staves, clapboards, shingles, boards, planks and timbers. Although Boston is still a large lumber market and has continued so through all these years, it did not long maintain its supremacy in this country, being early overshadowed by New York and many other markets, and now all of these are inferior to the great city of the West, Chicago.—*Timberman.*

IGNORANCE OR WASTE

To the Editor of the Canada Lumberman:

SIR,—A certain mill-owner, well-known to the writer, in reproving one of his employees, was met with the rejoinder beginning with, "I thought," but got no further, as he was promptly interrupted with, "You thought? Who told you to think? You have spoiled every piece in that pile. I want you to know that I am doing the thinking for this business, and if you do not do as I tell you, you will pay the cost of your thinking."

Without expressing an opinion upon the wisdom or disposition of the mill man, as shown above, I have often thought of the force of the sentiments expressed, when my business brings me into our country saw mills cutting hard woods. It is probably a safe assertion that ninety per cent. of the slabs other than pine, go to the wood pile without so much as a "thought" being expended upon them, but I came across an instance of thinking and doing, backed up with experience and figures, which may be of benefit to many a man, if the facts are understood.

SLAB SAWING

In one of the mills of Macpherson & Schell, of Alexandria, is a saw-table of special construction upon which is worked up the slabs and edgings into marketable shape.

The basswood slabs are cut into cigar box stock, 3-16 inch thick and of suitable widths and lengths, usually four feet long, and some into piling boards for rolling mills, trunk slats and other uses. Ash slabs and edgings were cut into wainscot lumber 7-8 inch thick, three and four inches wide and three and four feet long, and an examination of the finished stock showed a grain and surface not possible to equal from lumber from the body of the log. Birch and hard maple were cut into furniture stock, and soft maple into wainscot, making a fine white finish. For working up small second growth basswood into box boards, drawer stock and other furniture uses, the same firm have a miniature saw-mill, of their own special make, self-contained, easily removable if needed. We were informed that over two-thirds of the expense of operating the mill was cleared from the slab-sawing venture of the firm. Surely the above "experience" should cause many mill men to indulge in some thinking of a profitable nature, and if some of the "lumber merchants" would take up the matter with manufacturers, a more profitable trade awaits them than often is the case with larger operations.— WHITE BASSWOOD.

————

THE SO-CALLED waste stock is often the measure of profit or loss in a mill or factory.

————

IF A BELT persists in slipping after the machine is fairly under motion and is sufficiently tight, then it is evident that the pulley is deficient in frictional surface, being either too small in diameter or too narrow face; in either case it is far better economy to change the pulley than to go on purchasing new belts every few months.

SPLINTERS

A MODERN saw-mill is about as interesting a thing as one can see. The whole process of converting logs into lumber is laid bare, and there is a "go" about it, all of which is fascinating in the extreme.

IN BUYING wood-working machinery, it is better to get it a little heavier and stronger than the work intended for requires, rather than the reverse. A light machine can't be crowded without lengthening the repair account, and repairs cost not only money but time as well.

A POLICY on a steam saw-mill includes the whole machinery necessary to make it a saw-mill in all its parts, as well as the building.— *Bigler vs. New York Central Insurance Company*, 21 *Barb.* (N. Y.), 635 (1885). This case is affirmed in 22 N. Y., 402.

THE FOLLOWING, it is said, will fasten leather to iron or steel so firmly that they cannot be separated. Soak the leather with a warm solution of gallnuts, spread thinly over the metal a solution of the best glue (hot), place the two together with a pressure on them, and leave to dry.

————

TIMELY HINTS

IF THE iron wedge will not draw, build a fire of chips and heat it.

IF YOU place the axe near the stove for fifteen minutes it will cut better, and not be so apt to break along the edge.

HERE IS A hint which might be noted with profit by many concerns. The manager of a large southern company says: "By dressing and drying we reduce the weight of our lumber from 4,800 to 2,600 pounds per thousand, which gives us a big advantage in freight. Besides, we save insurance, rehandling and wharfage, and gain dispatch, which is oftentimes a big item in shipping by car.

SINCE IT HAS become a fact well established that steam pipes in contact with wood may cause a fire, wouldn't it be a good idea to surround such pipes with metal, something like you would a stove pipe that passes through a partition or floor?

EBONY WOOD weighs eighty-three pounds to the cubic foot; lignum vitæ, the same; hickory, fifty-two pounds; birch, forty-five pounds; beech, forty; yellow pine, thirty-eight; white pine, twenty-five; cork, fifteen, and water, sixty-two.

HEMLOCK is favorably considered for railroad ties, not especially for its durability, but for its property of holding spikes.

GOOD MACHINERY is a necessity in the sawmill, in the planing-mill, and in all wood-working establishments.

FEW PERSONS have any idea as to the amount of coal that can be stowed in a given space; we therefore give an example of the manner in which it may be figured up. A shed or room 15 feet high, 18 feet wide and 30 feet long will hold 200 tons of anthracite coal, and perhaps ten tons less of Cumberland. Thus 15 x 18 x 30 = 8,100 divided by 40. Average cubic contents of a ton of anthracite, $202\frac{1}{2}$.

IN THE HEATING of burns and scalds, where there is danger of contracting scars, rub the new skin several times a day with good sweet oil. Persist in this rubbing until the skin is soft and flexible.

To FIND the diameter of a pulley for any speed multiply diameter of pulley on main shaft by the revolutions (or speed) required, the quotient will be the diameter in inches of required pulley.

A SIPHON MOTOR is highly recommended for furnishing small power, especially in the country. Water can be readily siphoned from a running stream to where it is wanted. Then by the fall of the water from the outer leg of the siphon upon an overshot wheel the power may be obtained.

BUT ONE WAY.—There is but one way to get the full value of a machine, and that is to keep it in good repair, clean, well oiled and taken care of. Nothing ruins machinery like neglect

How About This?—Did it ever strike you, asks an exchange, that you may not be getting either the full quota of work from your machinery, or the best quality? Of course it is unwise to crowd a machine, but many a machine that is doing poor work, and perhaps little work, might be made to give better service, if well taken care of. Machinery may be made automatic; but there are no machines that will take absolute care of themselves. They respond to neglect and to attention, almost like sentient beings.

One thousand feet of rough white pine lumber when dry will weigh 2,500 pounds. Dress this lumber on one side and you reduce its weight to 2,200 pounds; dress it two sides and you reduce it to 2,000 pounds; work it into flooring you reduce it to 1,800 pounds; and work it into bevel siding and you reduce it to 1,600 pounds. Worth considering.

When I go into the woods in sharp, frosty weather I carry a few cotton rags in my pocket, and before driving an iron wedge into a frozen log I fold one of them across the point of the wedge. With this precaution there is no danger that the wedge will fly out, at a touch, as it is likely to without it.

Round Timber, when squared, is estimated to lose one-fifth.

FIFTY FEET OF BOARDS will build one rod of fence five boards high, first board being ten inches wide, second eight inches, third seven inches, fourth six inches, and fifth five inches.

———

NOTHING helps the introduction of a new machine or device among practical mechanics more than simplicity of design and the absence of numerous joints and pieces, which tend to shorten the life of the machine as well as impair its efficiency. Joints are good things to avoid where possible, as the inevitable wear is followed by lost motion, which affects the accuracy of the machine.—*Machinery.*

———

LUMBERMEN and all workers in wood, like agriculturists and the miners and manufacturers of the metals, are the world's real benefactors. They contribute more to the world's wealth than the followers of any other pursuit. The products they utilize are nature's gifts, whether it is the food men subsist upon or the clothes they wear, the tools they work with or the houses they dwell in. The produce of their labor is clear gain, and all the occupations are dependent upon them. There are no more useful members of the community than the men who fell the forest trees and fashion the wood into articles of utility.—*Nashville Southern Lumberman.*

SIBERIA'S TIMBER BELT

IT appears that Siberia, from the plain of the Obi river on the west to the valley of the Indighirka on the east, embracing the great plains, or river valleys, of the Yenisei, Olenek, Lena and Yana rivers, is one great timber belt, averaging more than 1,000 miles in breadth from north to south—being full 1,700 miles wide in the Yenisei district—and having a length from east to west of not less than 4,600 versts, about 3,000 miles. Unlike equatorial forests, the trees of the Siberian taigas are mainly conifers, comprising pines of several varieties, firs and larches. In the Yenisei, Lena and Olenek regions there are thousands of square miles where no human foot has ever been. The long-stemmed conifers rise to a height of 150 feet or more and stand so closely together that walking among them is difficult.

The dense, lofty tops exclude the pale Arctic sunshine, and the straight pale trunks, all looking exactly alike, so bewilder the eye in the obscurity that all sense of direction is lost. Even the most experienced trappers of sable dare not venture into the dense taigas without taking the precaution of "blazing" the trees constantly with hatchets as they walk forward. If lost there the hunter rarely finds his way out, but perishes miserably from starvation and cold. The natives avoid the taigas, and have a name for them which signifies "places where the mind is lost."—*The Canada Lumberman.*

A CURRENT item says: There is a general idea that beech timber or lumber has no especial place in the world and is of no practical use as a wood for building, in short that it is first-class for stove wood but useless for anything else. There are one or two markets in this country where they have found out that beech is actually good for something as a lumber wood. At these places you will be told that for heavy flooring, for factory and warehouse heavy floor timbers it has no superior. Also in these places which peculiarly try the wearing qualities of wood it will outlast anything else. For floors that require to hold up heavy machinery or heavy loads it is invaluable. "It may break but will never bend," or sag under weight.—*The Wood-Worker.*

GOOD ADVICE

NEVER take it for granted that a tool is in order. See that it is, before you set it and apply the power. A glance costs nothing, and it may detect an error or a defect, and thus save the labor of putting in and taking out unnecessarily. A first-class workman, in a first-class shop, will always have his tools in order, but a first-class workman takes nothing for granted that may be settled at once definitely by a touch of the finger or a glance of the eye.

Never take it for granted that a bearing is not hot, because it is not hot enough to make itself smelled all over the shop, nor that the water in the boiler is not dangerously low, because the boiler has not yet gone skyward through the roof. Never let any of the points be forgotten or wilfully overlooked. Forgetfulness and wilful neglect are at the bottom of 99 out of each 100 cases of "mysterious" fires, or breakages, or stoppages, or explosions.—*Northwestern Mechanic.*

BLOOD POISONING FROM MACHINE OIL

TAKE care, says *Power and Transmission*, how you let machine oil or lubricator come in contact with a cut or scratch on your hand or arm, as serious blood-poisoning may result. In the manufacture of some of these machine oils fat from diseased and decomposed animals is used. All physicians know how poisonous such matter is. The only safeguard is not to let any spot where the skin is broken be touched by any machine oil or lubricator.

WE HAD a man in our mill, who round a resaw lingered; he got his hand too near the teeth, and now he is unfingered.—*Northwestern Lumberman.*

TABLE OF DISTANCES AND TIME

Localities	Dist'e from N. Y	Time	The accompanying table shows the distance from the place named to New York City, by the usually travelled routes, generally by railroad; also the time at the same places when it is 12 o'clock, or mean noon, at New York.
	Miles	h. m.	
New York...	....	12.00	
Brooklyn...	....	12.00	
Montreal.....	401	11.58	
Boston.......	236	12.12	
Buffalo......	422	11.41	
Cleveland....	581	11.30	
Columbus....	650	11.24	
Cincinnati....	799	11.19	
Detroit......	663	11.24	
Indianapolis..	825	11.14	
Chicago......	868	11.06	
St. Louis.....	1087	10.55	
Omaha.......	1540	10.42	
Leavenworth.	1582	10.29	
Philadelphia..	88	11.56	
Baltimore....	185	11.50	
Pittsburg.....	431	11.36	
Louisville....	934	11.14	
Memphis.....	1072	10.54	
New Orleans .	1597	10.56	
Mobile.......	1448	11.05	
Savannah....	890	11.31	
Charleston....	794	11.36	
Richmond....	353	11.46	
San Francisco	3200	8.46	
Liverpool....	3000	7.16 P.M.	

INCREASE IN STRENGTH BY SEA=SONING LUMBER

Ash..........44.7% | Oak..........26.1%
Beech.........01.9% | White Pine..... 9 %
Elm..........12.3% |

INTEREST

INTEREST is a percentage paid for the use of money.

PRINCIPAL is the sum for the use of which interest is paid.

RATE PER CENT. is the sum paid on the hundred.

PER ANNUM means by the year.

AMOUNT is the principal and interest added together.

Rate per c.	Time in which a Sum will Double			
	Simple Interest		Compound Interest	
2	50 years		35 years	1 day.
2½	40 "		28 "	26 "
3	33 "	4 months	23 "	164 "
3½	28 "	208 days	20 "	54 "
4	25 "		17 "	246 "
4½	22 "	81 days.	15 "	273 "
5	20 "		15 "	75 "
6	16 "	8 months	14 "	327 "
7	14 "	104 days.	10 "	89 "
8	12½ "		9 "	2 "
9	11 "	40 days.	8 "	16 "
10	10 "		7 "	100 "

LEGAL RATES OF INTEREST

IN THE DIFFERENT STATES

MAINE, New Hampshire, Vermont, Massachusetts, Rhode Island, Connecticut, New York, Pennsylvania, Delaware, Maryland, Virginia, W. Virginia, North Carolina, Mississippi, Ohio, Indiana, Illinois, Iowa, Kentucky, Tennessee, Arkansas, Missouri, District of Columbia, Canada, New Brunswick, New Jersey, New Mexico, is 6 PER CENT.

South Carolina, Georgia, Michigan, Wisconsin, Minnesota, Dakota Territory, Kansas, is 7 PER CENT.

Alabama, Texas, Florida, is 8 PER CENT.

California, Oregon, Nebraska, Washington Territory, Nevada, Colorado, Montana, Idaho, Arizona, Utah, Wyoming, is 10 PER CENT.

Louisiana is 5 CER CENT.

A TABLE OF DAILY SAVINGS AT COMPOUND INTEREST

Cts. a Day	Per Year	In 10 Yrs.	Fifty Years
$ .02¾	$ 10.00	$ 130	$ 2,900
.05¼	20.00	260	5,800
.11	40.00	520	11,600
.27½	100.00	1,300	29,000
.55	200.00	2,600	58,000
1.10	400.00	5,200	116,000
1.37	500.00	6,500	145,000

BUSINESS LAW

IGNORANCE of the law excuses no one.

An agreement without consideration is void.

Signatures made with a lead pencil are good in law.

A receipt for money paid is not legally conclusive.

The acts of one partner bind all the others.

Contracts made on Sunday cannot be enforced.

A contract made with a minor or a lunatic is void.

Principals are responsible for the acts of their agents.

Agents are responsible to their principals for errors.

Each individual in a partnership is responsible for the whole amount of the debt of the firm.

A note given by a minor is void.

Notes bear interest only when so stated.

It is not legally necessary to say on a note "for value received."

A note drawn on Sunday is void.

A note obtained by fraud, or from a person in a state of intoxication, cannot be collected.

If a note be lost or stolen, it does not release the maker; he must pay it.

An endorser of a note is exempt from liability, if not served with notice of its dishonor within twenty-four hours of its non-payment.

It is fraud to conceal a fraud.

The law compels no one to do impossibilties.

A personal right of action dies with the person.

An oral agreement must be proved by evidence. A written agreement proves itself. The law prefers written to oral evidence, because of its precision.

MAXIMS

Gold goes in at any gate except heaven's.
Kind speeches comfort the heavy hearted.
He that blows in the dust fills his own eyes.
A quiet conscience sleeps in slumber.
Many are better known than trusted.
A light purse is a heavy curse.
The sickness of the body may prove the health of the soul.
By others' faults, wise men correct their own.
Simple diet makes healthy children.
It is a good horse that never stumbles.
Every man is architect to his own fortune.
The more a man talks the less he thinks.
Nothing venture, nothing have.
Beware of a silent dog and still water.
He that would thrive, must rise at five.
He that has thriven, may rise at seven.

———

SUBSTITUTE FOR BLACK WALNUT

In view of the growing scarcity of black walnut, black birch is largely taking its place, as well as that of cherry, which is also becoming very scarce. Birch has much the same color as cherry, and is just as easy to work as black walnut, and as suitable for nearly all the purposes for which that wood is used. When properly stained, it is nearly impossible to distinguish it from walnut, and it is susceptible of a beautiful polish, equal to that of any wood used in the manufacture of furniture. Large quantities of it are imported from Canada, in some parts of which it is very plentiful and cheap, costing only about a dollar per hundred feet at the saw-mills.

THE ——

HORSEMAN'S FRIEND

BY

PROF. JAMES LAW, V.S.

A complete and handy treatise on Domestic Animals with an article by J. G. Rutherford, Chief Veterinary Inspector, on the Breeding in Canada of Horses for Army use.

NO FARMER SHOULD BE WITHOUT IT

500 pp., Cloth **$1.00**

For Sale by all Booksellers

PUBLISHED BY

THE MUSSON BOOK CO. LIMITED

TORONTO

Publications by Algrove Publishing Limited

If you like this book, the following list of other titles from our popular *"Classic Reprint Series"* may be of interest to you.

ALGROVE PUBLISHING LIMITED

36 Mill Street, P.O. Box 1238, Almonte, Ontario, Canada K0A 1A0
Tel: (613) 256-0350 Fax: (613) 256-0360 Email: sales@algrove.com